Laboratory Manual
Electrical Engineering
Uncovered

Dick White
Roger Doering

Prentice Hall
Upper Saddle River, New Jersey 07458

Acquisitions editor: Eric Svendsen
Editorial assistant: Andrea Au

Laboratory Manual for Electrical Engineering Uncovered, 1997

©1997 Prentice-Hall, Inc.
Simon & Schuster / A Viacom Company
Upper Saddle River, New Jersey 07458

Printed in the United States of America

10 9 8 7 6 5 4 3 2 1

ISBN 0-13-712622-0

Prentice-Hall International (UK) Limited, *London*
Prentice-Hall of Australia Pty. Limited, *Sydney*
Prentice-Hall Canada Inc., *Toronto*
Prentice-Hall Hispanoamericana, S. A., *Mexico*
Prentice-Hall of India Private Limited, *New Delhi*
Prentice-Hall of Japan, Inc., *Tokyo*
Simon & Schuster Asia Pte. Ltd., *Singapore*
Editora Prentice-Hall do Brasil, Ltda., *Rio de Janeiro*

Laboratory Experiments

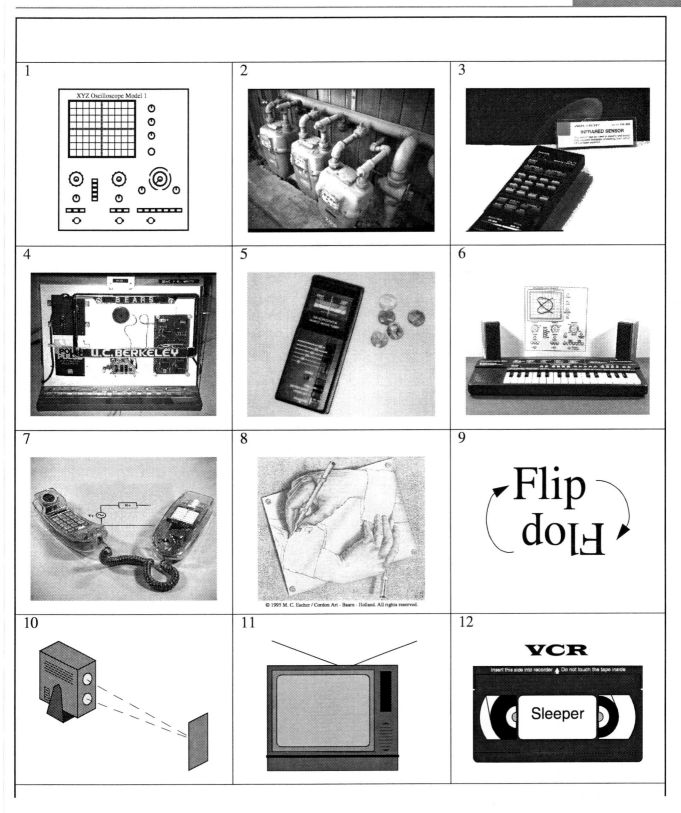

1

2

3

4

5

6

7

8

© 1995 M. C. Escher / Cordon Art - Baarn - Holland. All rights reserved.

9

Flip Flop

10

11

12

VCR

Insert this side into recorder ▲ Do not touch the tape inside

Sleeper

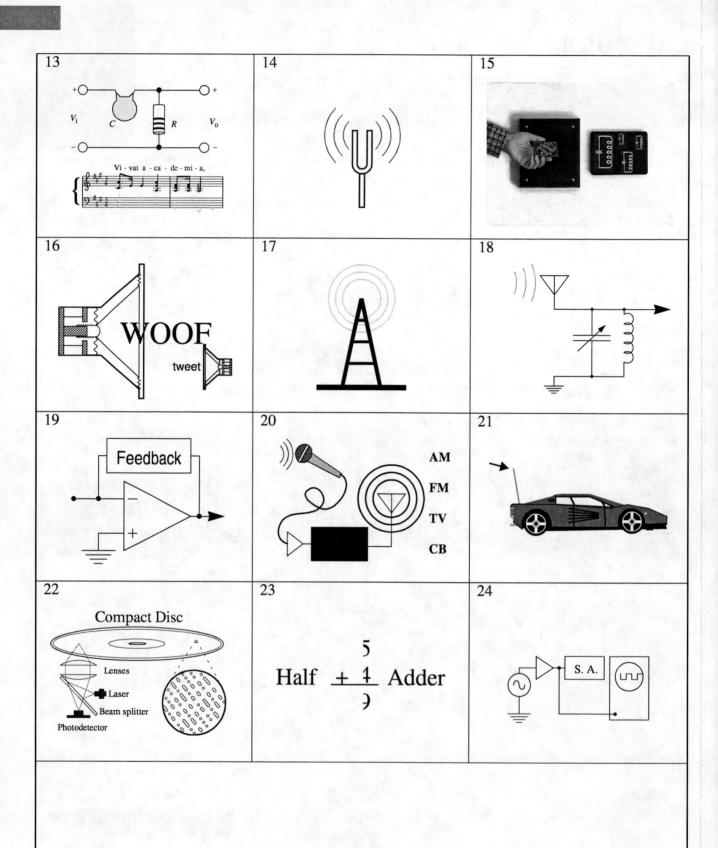

Contents

continued...

Contents

Preface

These laboratory experiments probably involve electronic equipment that you've already used but perhaps not understood — such as a VCR, CD player, remote control for TV, and an instant-camera ultrasonic focussing unit.

These experiments were developed for use in an introductory college electrical engineering course in which students are encouraged to develop intuitive ideas about how electronic devices and systems work. The name of the accompanying text, *Electrical Engineering Uncovered,* actually comes from the lab, where most pieces of equipment have a transparent cover (or perhaps no cover at all), so that one can see inside. We have only one of most lab set-ups, and no more than twelve students can work at a time in the lab, in two-person teams, on these two-hour experiments. We aim for an atmosphere rather like that in an industrial lab where many pieces of equipment are unique. In addition, students are required to have the summary pages of their lab manual witnessed, and the importance of cost is emphasized by the presence of price tags on most equipment. To encourage preparation, for each lab there is a tear-out pre-lab question sheet that is to be completed and handed to the instructor at the start of each lab.

The list of labs that follows is arranged hierarchically, meaning that once you've completed a given lab, you can do those indented under it. There are more experiments here than a given student will complete in a ten- or fifteen-week term. Students can therefore select some of the later labs on the basis of their own interests. On occasion, some students create a new lab, which they find very motivating, and which adds to our stock of labs.

We hope that you enjoy this material while learning from it. And if you find omissions or mistakes, or develop additional labs, please get in touch.

Dick White and Roger Doering
Electrical Engineering and Computer Sciences Department
University of California, Berkeley, CA 94720

Lab Project List

This list is organized hierarchically: once you've completed a given lab, you can do those indented under it.

Oscilloscope Familiarization
Parallel and Series Components
    Infrared Remote Control
        Television
            VCR
    Ultrasonic Rangefinder
    Guitar Tuner & Speaker Response
        Lissajous Figures
            Touch-Tone Telephone
Curve Tracer
Astable Multivibrator
    Infrared Door Alarm
RC Filters
    Resonant Filters
    Card Key
    Loudspeaker Crossover
AM Radio Transmitter
    AM Radio Receiver
    Op-amps
        Multiband Radio Transmitter
            Remote Control Car
CD Player
Half Adder
Simulink

How Do You Work in a Laboratory?

Establishing truth in engineering involves doing a successful experiment. The experiments that follow are designed so that you can see clearly what happens and reach unambiguous conclusions about the subject matter at the end of each experiment. Incidentally, we've tried to make the lab experience rather like that in an exploratory industrial lab. Here, briefly, are the elements.

1. Preparing. In order to finish each experiment in just two hours you'll need to look at the subject matter before you come to lab. You should also reserve the next experiment that you want to do, since there's only one set-up for most of the experiments. Read over the lab notes before you come into the lab and bring with you the completed pre-lab quiz.

2. Satisfying your curiosity. Almost every piece of equipment has a transparent case, so you can see what happens inside. Take a look at the equipment you'll be using in today's lab. Sometime, look at the other equipment in the lab, to see what else is there that you might find interesting. Take a look at the written materials in the lab — such as David Macaulay's *The Way Things Work*, electronics parts catalogs, the Edmund Scientific Company catalog, the *Radio Amateur's Handbook*, the small booklets from RadioShack about particular topics in electronics, and the charts on the walls. And notice the price tags on the equipment and try to figure out what factors might determine the widely different prices.

3. Tinkering. Years ago, people who selected engineering as a career had done a lot of tinkering as youngsters. It seems that far fewer engineering students today have had that experience. This lab is a place where you can try out some things that aren't required. But check with the lab instructor before trying anything that might be hazardous.

4. Observing. Sometimes people have discovered important new phenomena by noticing an unexpected small change on an oscilloscope trace or meter indication. Get in the habit of observing closely in the lab.

5. Working together. As in most of industry, teamwork in this lab is encouraged — in fact, it's required! If you're on a team, you have other people to "bounce ideas off", and there's more than one source of ideas. Two-person teams are best but both of you should contribute equally.

6. Taking data once. Take data directly on your lab sheets. Don't waste time prettying up your data. If the lab requires graph paper to plot data points, you may use (or copy) one of the following pages of graph paper.

7. Drawing conclusions. The lab sheets ask you to state your conclusions before you leave the lab. Savvy students think about what conclusions are likely before coming into the lab, during the lab preparation stage.

8. Having ideas and inventing. People invent all the time, particularly if they are by habit observant and thoughtful. Past students in this course have come up with interesting ideas that the instructors hadn't thought of. Let your lab instructor know.

9. Making a lab book a legal document. Common practice in industry is to sign and date every page of your lab notebook, and to have another worker witness and date the pages. You're asked to do a version of that here in order to develop the habit.

10. Learning from your lab instructor. Chat with the lab instructor. Ask questions. Try out interesting ideas that you get. And you might want to find out about the lab instructor's own interests and goals.

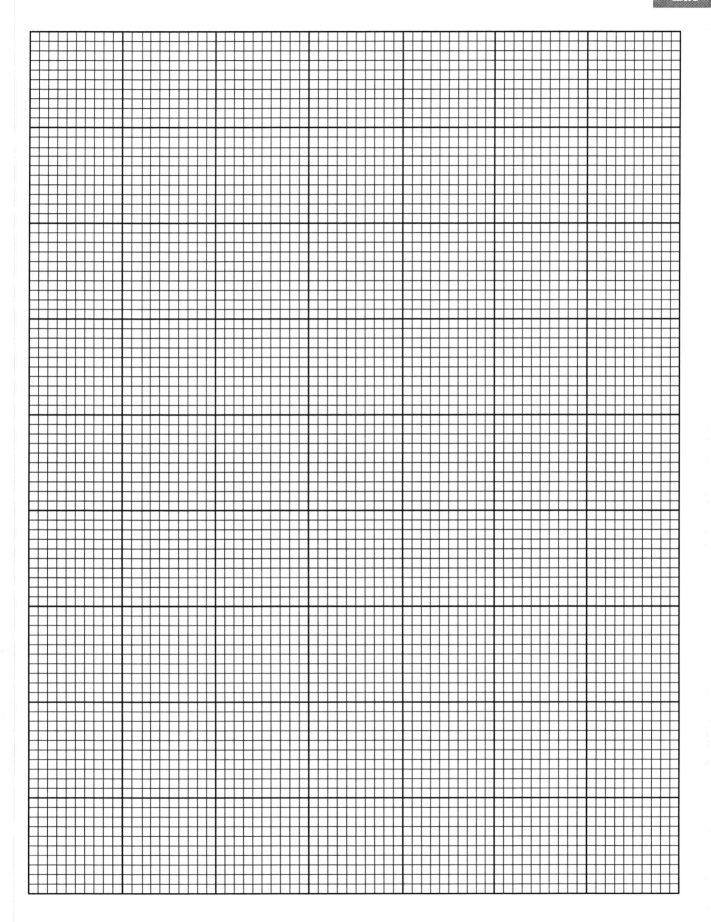

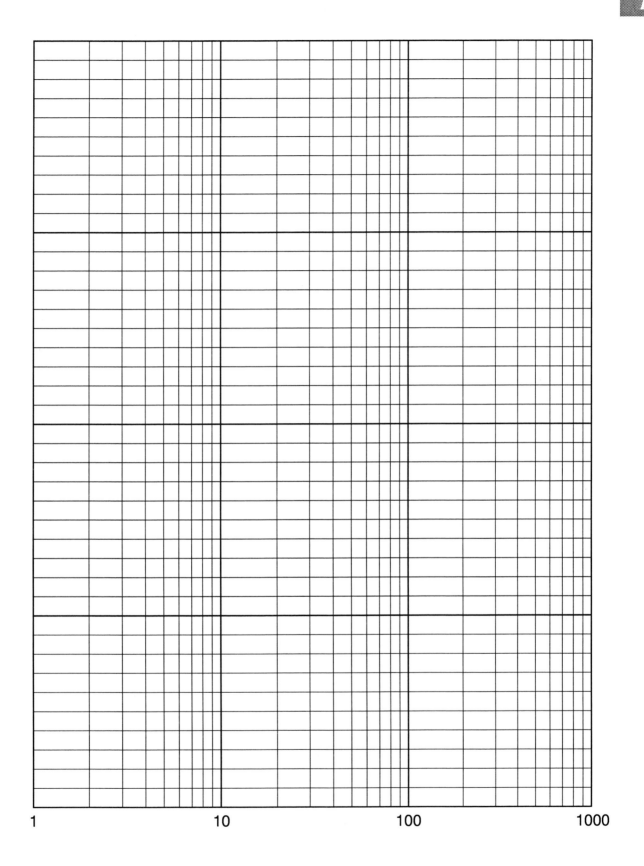

1   10   100   1000

Semi-log

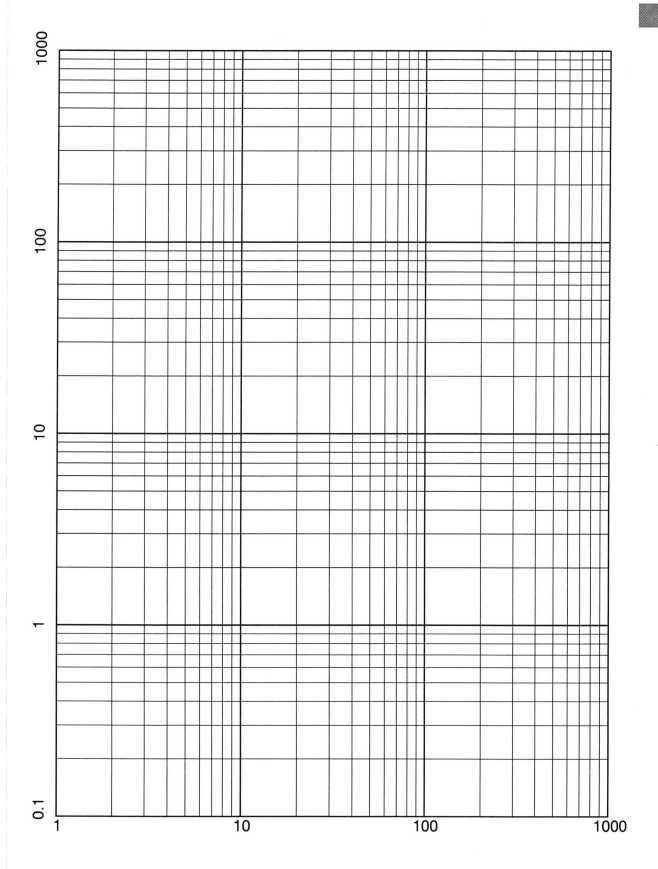

Log vs. Log

Custom Lab

Lab 0

Instructional Objectives (At the end of this lab you should be able to:)

I1.

_____
_____
_____

Description and Background

_____
_____
_____
_____
_____

Equipment

_____
_____
_____

Procedures

P1.

_____
_____
_____
_____
_____

P2.

_____
_____
_____
_____
_____

References

Dick White and Roger Doering, *Electrical Engineering Uncovered* (Upper Saddle River, N.J.: Prentice Hall, 1997)

_____
_____

Custom Lab

Questions

  Q1.

Conclusions (What did you learn from this experiment?)

    _____

    _____

    _____

Data and Observations

Signature: _____ Date: ___/___/___ Witness: _____ Date: ___/___/___

Prelab Questions: Custom Lab

(Bring sheet with questions answered to your lab session)

Print your name (Last, First): _____

Q1.

_____
_____
_____
_____
_____
_____
_____

Oscilloscope

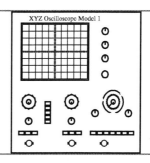

Lab

1

Instructional Objectives (At the end of this lab you should be able to:)

I1. Identify the functions of the major controls of an oscilloscope.

I2. Display a periodic waveform stably and measure characteristics of it such as amplitude and frequency.

I3. Use an oscilloscope probe and predict its effect on the amplitude of a signal.

Equipment

Common oscilloscope (preferably analog); *Science Fair 130-in-one Electronic Project Lab*; signal source providing sine waves of variable frequency; scope probe; coaxial cables to connect scope and signal source; 1.5-volt battery; cliplead-to-coaxial adaptor; scope camera and instant film (if available); two-inch length of bare wire

---

Description And Background

The oscilloscope ("scope") is a most versatile and useful piece of electronic test equipment. Effective use of a scope is an essential skill for the student of electronics.

In the scope's <u>display</u> (cathode-ray) tube, an electron beam is deflected by electric fields across the fluorescent front surface to produce a visible display. Scopes typically have two <u>input</u> ("vertical") <u>channels</u> and switches that permit one to view up to two waveforms (voltage that varies with time) to be displayed simultaneously ("chop") or alternately ("alt"). When an input is in the "DC" mode, a scope can measure a steady voltage; in "AC" mode, a scope will display only the time-varying component of a signal. The <u>timebase</u> section of a scope produces voltages that deflect ("sweep") the electron beam in the horizontal direction at speeds that can range from centimeters per second up to centimeters per nanosecond. By proper adjustment of the "triggering" controls, the start of the horizontal deflection signal can be synchronized to a feature of an input signal under study so as to produce a stable display. To display a repetitive waveform, a repetitive sweep is used; a single-sweep setting is used when displaying a signal that occurs only once. Some scopes provide a "delayed sweep" for displaying a portion of a waveform at a higher resolution or sweep speed. Another capability of many scopes that is exploited in the lab called "Lissajous 'Scope Patterns" is that of driving both the horizontal and the vertical deflections with time-varying signals that you supply in order to determine the relative frequencies of the two signals.

The least expensive scopes may be based on analog principles and be able to display signals whose frequency content ranges from 0 Hz (DC) up to tens or a few hundred MHz. Digital scopes are more costly but can capture and display non-repetitive waveforms; they may also store and send to a hardcopy printer representations of such waveforms. Beware though, that digital scopes can hide waveform features you may need to see.

Procedures

P1. Select any of the sound producing projects in the entertainment section of the RadioShack *Science Fair 130-in-one Electronic Project Lab*. Build the circuit to use as a signal source. Draw the schematic.

P2. To see whether your scope is working, insert the short length of bare wire (or an unbent paper clip) into the center of the BNC input connection on your scope and hold onto the wire with your fingers while your other hand is free of contact with anything metallic. You should see a 60 Hz waveform on the screen. Determine that the frequency is indeed 60Hz from the pattern on the screen and the setting of the timebase switch. Determine the amplitude of the waveform (it may be several volts). Speculate on the origin of this "signal".

P3. Measure the voltage across a battery. Try it with DC and then AC input coupling.

P4. Connect the signal source to the input terminal of a channel of the scope using a scope probe. Adjust the amplitude and the timebase controls until the waveform is stable and entirely on the screen. Sketch the waveform. Measure and record the waveform amplitude in volts and the period of the waveform (period is the duration of time for the waveform to repeat). If the waveform is periodic, find and record the frequency of the waveform (frequency in Hz is the reciprocal of the period in seconds).

P5. Repeat P4 with a coaxial cable instead of the scope probe.

References

P. Horowitz and W. Hill, *The Art of Electronics*, 2nd ed. (Cambridge, England: Cambridge University Press, 1989), Appendix A, pp. 1045-49; pp. 55-6; 57; 57-8; 879.
David Macaulay, *The Way Things Work* (Boston, MA: Houghton Mifflin, 1988), 262

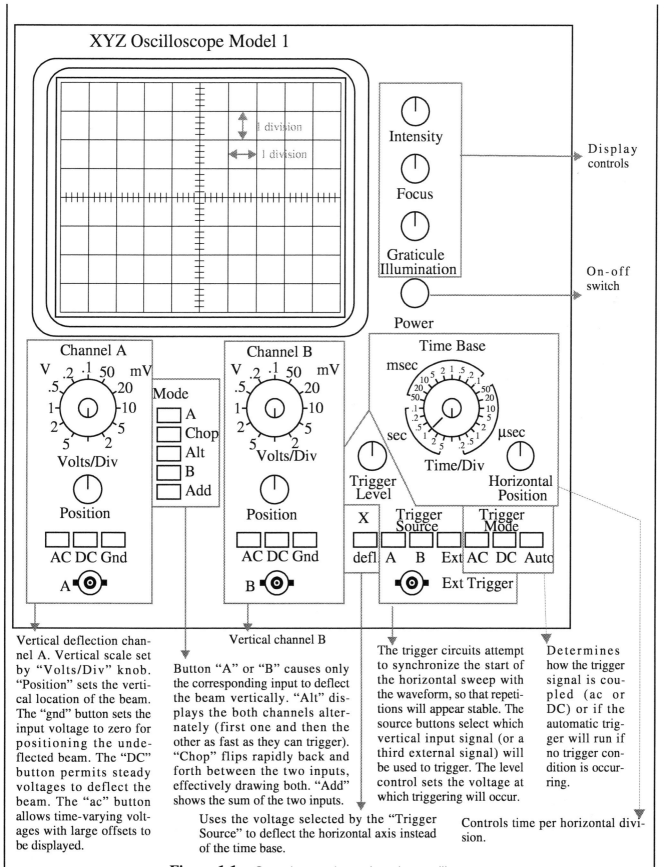

Figure 1.1. Generic two-channel analog oscilloscope.

Vertical deflection channel A. Vertical scale set by "Volts/Div" knob. "Position" sets the vertical location of the beam. The "gnd" button sets the input voltage to zero for positioning the undeflected beam. The "DC" button permits steady voltages to deflect the beam. The "ac" button allows time-varying voltages with large offsets to be displayed.

Vertical channel B

Button "A" or "B" causes only the corresponding input to deflect the beam vertically. "Alt" displays the both channels alternately (first one and then the other as fast as they can trigger). "Chop" flips rapidly back and forth between the two inputs, effectively drawing both. "Add" shows the sum of the two inputs.

Uses the voltage selected by the "Trigger Source" to deflect the horizontal axis instead of the time base.

The trigger circuits attempt to synchronize the start of the horizontal sweep with the waveform, so that repetitions will appear stable. The source buttons select which vertical input signal (or a third external signal) will be used to trigger. The level control sets the voltage at which triggering will occur.

Determines how the trigger signal is coupled (ac or DC) or if the automatic trigger will run if no trigger condition is occurring.

Controls time per horizontal division.

Oscilloscope

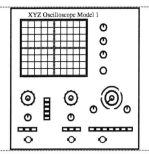

XYZ Oscilloscope Model 1

Questions

Q1. Where should the scope probe ground clip be connected?

_____

Q2. Speculate on the origin of the 60Hz waveform you observed (procedure P2).

_____

Q3. Why should we use a scope probe instead of a coaxial wire?

_____

Conclusions (What did you learn from this experiment?)

_____

_____

_____

Data and Observations

Signature: _____ Date: ___/___/___ Witness: _____ Date: ___/___/___

Prelab Questions: Oscilloscope

(Bring sheet with questions answered to your lab session)

Print your name (Last, First): _____

Q1. Explain how an oscilloscope draws on the screen.

_____
_____
_____
_____
_____
_____
_____
_____

Q2. What is meant by "triggering"?

_____
_____
_____
_____
_____
_____
_____
_____

Series and Parallel

Components

Instructional Objectives (At the end of this experiment you should be able to:)

I1. Identify series and parallel connections of circuit components.

I2. Calculate (given the values of the circuit components) the total resistance (capacitance) of series- and parallel-connected resistors (capacitors).

I3. Estimate intuitively for a simple network of series- or parallel-connected resistors (capacitors) which resistor (capacitor) will dominate or carry the larger current.

Equipment

Science Fair 130-in-one Electronic Project Lab; multimeter; capacitance meter (if available).

Description and Background

Resistors or capacitors may be connected in series or parallel for many reasons, such as to reduce a voltage to a convenient value (as in a voltage divider) or to provide a value different from one that is commercially available. You therefore need to be able to find the effective resistance or capacitance of such a combination of individual elements. Handy test for parallel connection: where two or more components connect, if the current (or water or electricity) has a choice of which way to go (pipe or wire), then the components are connected in parallel there.

WATER MODELS

Since electricity can't usually be seen, for purposes of visualization it is convenient to consider <u>water models</u> for electrical quantities (see *Electrical Engineering Uncovered* for more discussion of these water models). Electric current flowing in a wire is analogous to the flow of water in a pipe. The voltage across a circuit element — the potential difference between the two ends of the element in a circuit — is analogous to a difference of water pressure between two points in a network of plumbing. The pressure difference might be produced by a pump driven by a constant-torque motor (Figure 2.1). The water models for a resistor, capacitor and inductor are shown in Figure 2.2.

1. Series and Parallel Resistors

Two "water resistors" connected in series are shown at the left in Figure 2.3. It should be clear that the same amount of water ("current") flows through each resistor, and that the pressure difference between A and B ("V_{AB}") is the sum of the pressure drops in resistors 1 and 2. For two water resistors connected in parallel (at the right in Figure 2.3) it should be clear that the total flow equals the sum of the flows in the two resistors, and that both resistors have the same pressure drop across them.

2. Series and Parallel Capacitors

The capacitor water model in Figure 2.2 is a close-fitting piston in a water-filled pipe fitted with a spring. To push more water in from the left (like adding charge to a "real" capacitor) requires more and more force — an increasing pressure difference between A and B — just as the voltage across a capacitor increases as it is charged.

For two capacitors in series (at left in Figure 2.4) we can see that the amount of water (analogous to charge) added to the left of the piston is the same in both capacitors, and the pressure difference from A to D is the sum of the pressure drops across each capacitor. You are left to draw your own conclusions about the parallel-connected water capacitors (at right of Figure 2.4).

Procedures

P1. Measure the values of all of the resistors in the Kit (record values under "**Data And Observations**" below). Answer question Q1 under "**Questions**" below.

P2. Connect three various pairs of resistors in series and then parallel, and measure the resistance of the combinations.

P3. Derive the formulas for series and parallel connected resistors from Ohm's Law ($V = IR$). (Put derivations on the back of the report.)

P4. Repeat P1 for capacitors.

P5. Repeat P2 for capacitors.

P6. Derive the formulas for the total capacitance of series-and parallel-connected capacitors. Hints: Use the formula for the capacitance of a parallel-plate capacitor: $C = \varepsilon$ (permittivity of dielectric) x A (area of plates) / d (distance separating plates). For series-connected capacitors, note that the same current flows through each capacitor. Hold one dimension constant in each case — area for series and separation for parallel. Assume a constant dielectric material.

P7. Build and test the two circuits in the Science Fair 130-in-one Electronic Project Lab that demonstrate these principles.

References

P. Horowitz and W. Hill, *The Art of Electronics*, 2nd ed. (Cambridge, England: Cambridge University Press, 1989), pp. 6; 21.

David Macaulay, *The Way Things Work* (Boston, MA: Houghton Mifflin, 1988), pp. 286-7.

Science Fair 130-in-one Electronic Project Lab (Fort Worth, TX: RadioShack [Tandy Corp], 1989), p. 31, 32.

Dick White and Roger Doering, *Electrical Engineering Uncovered* (Upper Saddle River, N.J.: Prentice Hall, 1997), "What Can You Do with These Components?" on page 193.

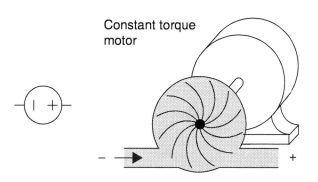

Figure 2.1 Water model for a steady voltage source (schematic symbol at left)

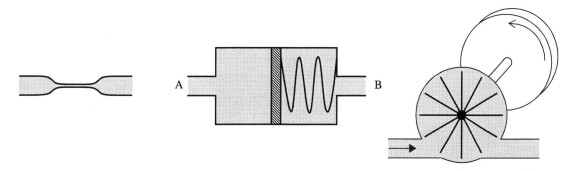

Figure 2.2 Water models and schematics symbols for a resistor (left), capacitor (center) and an inductor (right)

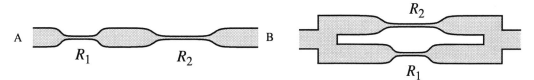

Figure 2.3 Water models for resistors connected in series (left) and parallel (right)

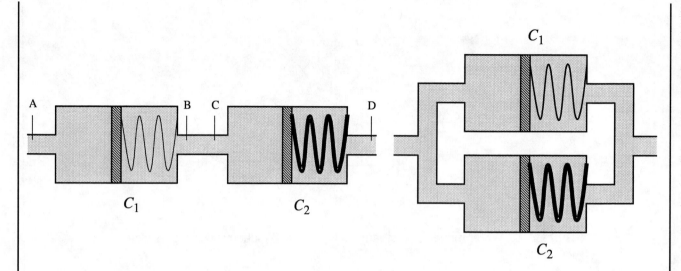

Figure 2.4 Water models for capacitors connected in series (left) and parallel (right). Capacitor C_2 has a stiffer spring than capacitor C_1.

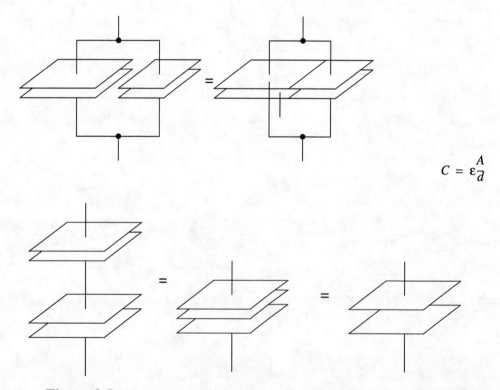

$$C = \varepsilon \frac{A}{d}$$

Figure 2.5 Parallel and series capacitor thought experiments for use with P6.

Series and Parallel

Components

Questions

Q1. Why are resistor values not precisely correct (why do measured resistance values differ from what the color code indicates)? By what percentage are they in error? Note: Percentage error = 100 x [Indicated value - Measured value] / [Indicated value].

_____

Conclusions (What did you learn from this experiment?)

_____

_____

_____

Data and Observations

Signature: _____ Date: ___/___/___ Witness: _____ Date: ___/___/___

Prelab Questions: Series and Parallel Components

(Bring sheet with questions answered to your lab session)

Print your name (Last, First): _____

Q1. What quantity (voltage, current or charge) is the same when you connect two different components as indicated below and apply a voltage across the terminals:

a. Two resistors in series?

b. Two resistors in parallel?

c. Two capacitors in series?

d. Two capacitors in parallel?

Infrared Remote Control

Lab
3

Instructional Objectives (At the end of this lab you should be able to:)

I1. Describe how infrared remote controls operate.

I2. Explain the advantage of using a carrier signal.

Equipment

Storage oscilloscope; infrared remote controls; infrared light sensor strip; infrared phototransistor and LED; two Science Fair 130-in-one Electronic Project Labs.

Description And Background

In this experiment you will use *infrared* (IR) radiation produced by a remote control device, and by a circuit that you build using a voltage source and an infrared light-emitting diode (LED). The infrared portion of the optical spectrum is characterized by wavelengths ranging from roughly one micron to a few hundred microns. Thus, the wavelength of infrared radiation is larger than the wavelength of red visible light. The human eye cannot detect infrared radiation directly; in this experiment you'll detect IR radiation with a special sensor strip that emits visible light when irradiated with IR, and with a phototransistor that produces a current when irradiated.

Infrared remote controls are very commonly used to control television sets, VCRs, cable converters, and stereos. In this lab we will use a phototransistor to convert the infrared light signal to an electrical current so that we can observe the signal on an oscilloscope.

Phototransistors operate by allowing a current to pass from their collector to their emitter that is proportional to the amount of light that strikes them. Infrared phototransistors are encased in a dark plastic that filters out the unwanted visible light and allows the infrared light to reach the transistor. They have only the two leads as the light takes the place of the base current. They will work well only if connected properly. If yours doesn't work try reversing the leads. The phototransistors are sensitive to light coming into the domed surface, so take care to align the remote with it.

A battery on the kit board will provide the voltage so that a current may flow through the phototransistor. Since the oscilloscope doesn't measure current we will need to add a resistor to the circuit to convert the current into a measurable voltage. While you might think that a large value resistor would work well for this, since it should give larger voltage changes for a given size current change, in fact you will get the best results with a 100-ohm resistor because of the internal capacitance of the phototransistor.

Infrared Light-Emitting Diodes (LEDs) convert a current to a very narrow band of light in the 950 nm wavelength region. Please be sure to connect a resistor in series with the LED. Because they are diodes, they will only allow current to flow in one direction. If you connect it up backwards, nothing will happen.

In the Procedures section the word "design" appears. Try not to let this frighten you. We're only talking about figuring out how to connect three or four components together with an equal number of wires.

Procedures

P1. Use the infrared sensor strip to be sure that your remote control is working.

P2. Design and construct a simple infrared (IR) detector circuit. Connect a storage scope and capture the waveforms from the remote. The infrared phototransistor will allow a current to pass through it when struck by infrared light. You will, of course, need a voltage source to push current through the phototransistor. Since the oscilloscope measures voltage, not current, you will need to convert the current to a voltage with a component. It is black in appearance because the plastic is a filter to block the visible light.

P3. Measure the output of different keys on the remote. Record the binary codes for at least the number keys. Compare the codes with those of a different remote from the same company, then a different company. Beware: many remote controls transmit a "same as last key" code when you hold a button down.

P4. Design and build your own simple IR remote to send signals as far as you can. **Do not to apply a voltage directly across the infrared LED**. They operate with 10 to 20 ma of *current*. A voltage source, like a battery, will burn them out instantaneously. Use the telegraph key switch to signal with. Use the infrared sensor strip to be sure that your circuit is working.

P5. Measure the remote output on the back of the CD player (you may need to share the CD player). What's different? Why?

References

P. Horowitz and W. Hill, *The Art of Electronics*, 2nd ed. (Cambridge, England: Cambridge University Press, 1989), pp. 57, 996.

David Macaulay, *The Way Things Work* (Boston, MA: Houghton Mifflin, 1988), pp. 292-293.

Infrared Remote Control

Questions

Q1. Why are the high and low parts of the waveform different from each other?

_____

Q2. What is the frequency of the carrier signal? _____

Q3. Why is this carrier signal used (very important!)?

_____

Conclusions (What did you learn from this experiment?)

_____

_____

_____

Data and Observations

Signature: _____ Date: ___/___/___ Witness: _____ Date: ___/___/___

Prelab Questions: Infrared Remote Control

(Bring sheet with questions answered to your lab session)

Print your name (Last, First): _____

Q1. Why do you suppose infrared radiation is used in remote tuners instead of, say, visible light or a radio signal?

Ultrasonic Rangefinder

Instructional Objectives (At the end of this lab you should be able to:)

I1. Relate frequency, velocity and wavelength.

I2. Describe how the transducers used in this experiment work.

Equipment

Ultrasonic measuring device (either the Micronta Electronic Tape Measure with Memory, Model 63-645), or the Polaroid Rangefinder Evaluation Kit); a tape measure with metric and english scales; piezoelectric tweeter (to use as an ultrasonic microphone); dual-timebase oscilloscope (a storage or digital oscilloscope is helpful if the measuring device doesn't provide a repetitive signal); various objects from around the lab to use as targets to detect; and a sweater (use it as an absorber to see how the rangefinder responds)

Description and Background

In the early 1970s, the Polaroid Corporation developed a distance measuring transducer for its SX-70 instant camera. This sensor had to be inexpensive, yet reasonably accurate and operate quickly enough to focus the camera.

Polaroid subsequently marketed the device to others for use in a wide range of applications. In the lab you will make use of one of these devices to make distance measurements, and to discover how it operates and what its limitations are.

Any transducer is a device that converts energy between one form and another. Most such devices work bidirectionally, converting between the two forms of energy.

Procedures

P1. Make a few trial measurements to determine the units of measure and operating procedures.

P2. By making experimental measurements, determine the device's limitations including minimum and maximum distances, beam shape (how far off axis can an object be seen vs. distance from the device?), smallest detectable object, and sensitivity to angles of beam incidence.

P3. Position the tweeter adjacent to the transducer on the measuring device and connect it to the vertical input on the oscilloscope using a 50 ohm coaxial cable. Set the scope input to 50 ohms. Measure the pulse duration, repetition rate (if any) and the frequency of the ultrasound.

P4. Inspect the transducer closely. It is not a piezoelectric device like the tweeter, but rather a capacitor with one fixed (rear) and one moveable (front) plate.

References

David Macaulay, *The Way Things Work* (Boston, MA: Houghton Mifflin, 1988), pp. 318-319, 372.

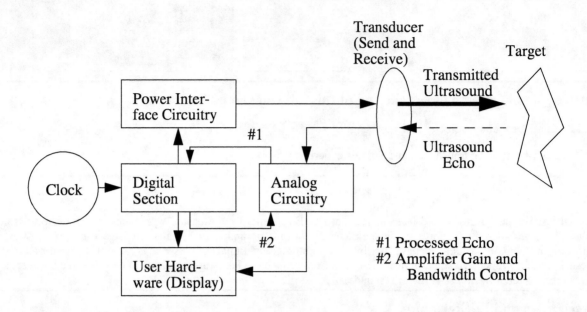

Figure 4.1 Block diagram — transmitting / receiving

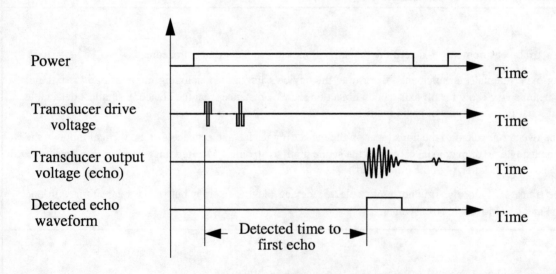

Figure 4.2 Waveforms

Ultrasonic Rangefinder

Questions

Q1. What is the frequency of the ultrasound?_____. What is its wavelength?_____ Why do you suppose the engineers selected this frequency?

_____

_____

Q2. Why does the transducer require a large bias voltage be placed across it?

_____

_____

Q3. What fundamental quantity is being measured to determine the distance?_____. Give the formula used to convert this quantity to distance.

_____

Q4. Since the frequency is out of our hearing range, why can we hear a click?

_____

Conclusions (What did you learn from this experiment?)

_____

_____

_____

Data and Observations

Signature: _____ Date: ___/___/___ Witness: _____ Date: ___/___/___

Prelab Questions: Ultrasonic Rangefinder

(Bring sheet with questions answered to your lab session)

Print your name (Last, First): _____

Q1. Before the lab, look up and record here the speed of sound in air, including the units and the conditions under which that applies (temperature, pressure, etc.).

Speed of sound in air _____ under these conditions: _____

Q2. The relationship among the frequency, wavelength and velocity of any wave is just

$$f \text{ (frequency)} \times \lambda \text{ (wavelength)} = v \text{ (velocity)}.$$

If the operating frequency of an ultrasonic motion detector is 25 kHz, what is the approximate wavelength of the wave it produces in air?

Guitar Tuner and

Loudspeaker Response

Instructional Objectives (At the end of this lab you should be able to:).

I1. Use semi-log graph paper.

I2. Calculate the frequency of any musical note.

I3. Describe what is meant by the frequency spectrum.

Equipment

Audio signal generator; guitar tuner; speaker; sound level meter (optional: a frequency counter).

Description And Background

TONES SOUNDED SINGLY: THE NAMES AND FREQUENCIES OF MUSICAL NOTES:

Many centuries were required for the evolution of the "Western music" of North America and Europe. (Since the electronic equipment used in the lab experiment is based on the Western musical conventions, we won't discuss here the musical scales and conventions of other cultures.)

We will take for our starting point the guitar fretboard sketched below.

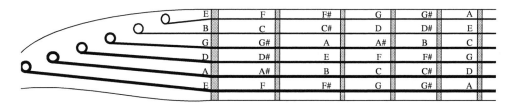

E	F	F#	G	G#	A
B	C	C#	D	D#	E
G	G#	A	A#	B	C
D	D#	E	F	F#	G
A	A#	B	C	C#	D
E	F	F#	G	G#	A

Figure 5.1 A stylized fretboard arranged as you might look at it from the playing position

The whole step pitches are identified by letters of the alphabet that run from A to G. Half step pitches (frequencies), which lie between those of the whole steps that bracket them, are denoted by their relations to the pitches on either side of them. The symbols used to denote these relations are: ♯ ("sharp") to denote a higher pitch, and ♭ ("flat") to denote a lower pitch. Note that two intervals B-C and E-F lack intervening tones. These two intervals are actually half steps, a feature of the western musical scales.

The <u>absolute</u> frequencies — values in Hertz — are set by convention: commonly the "middle A" note has a frequency of 440 Hz (it is a sound wave that varies sinusoidally 440 times per second). The <u>relative</u> frequencies of the notes of the scale are determined basically by "what tones sound good together".

Guitars use six strings tuned at fixed pitches and a series of frets to shorten the strings to allow other notes to be sounded. The basic six notes can be tuned relative to each other, which is fine for playing the guitar solo, but if the instrument is to be used with other instruments, an absolute frequency reference will allow all the instruments to harmonize. The guitar tuner is a such a reference.

The low string on the guitar is the fattest one and produces the lowest pitch. The note names of the strings proceeding from lowest to highest are E, A, D, G, B and E. Each interval is five half steps except the one from G to B which is four. This means the total number of half steps from the top string to the bottom is 24. This interval covers two octaves (the frequency of a note one octave higher than the other has twice the frequency of the other).

Procedures

P1. Use an audio signal generator and an inexpensive speaker to find the frequencies of the six guitar strings using the electronic tuner. Beware of false nulls (the needle jumps through zero). Plot the frequencies on the special graph provided. Answer question #1.

P2. Using the fact that the frequency doubles between octaves, derive a constant M, corresponding to one halftone, which when multiplied by the frequency of any pitch will give the frequency of the next higher pitch: freq(C)*M = freq(C♯)

P3. The tuner reads in the relative units called "cents". Make measurements to figure out what a "cent" is (caution: don't use 50-cent measures as the meter is most accurate near the middle of its range). Be sure to look at frequency ratios, not differences.

P4. Measure the sound pressure level at a fixed distance (state it!), and a fixed output level on the generator. Locate all the local maxima and minima (peaks and valleys) between 100 and 1000 Hz. Plot your results.

References

A. H. Benade, *Fundamentals of Musical Acoustics*, 2nd Revised Ed. (New York: Dover Publications, 1990).

P. Horowitz and W. Hill, *The Art of Electronics*, 2nd ed. (Cambridge, England: Cambridge University Press, 1989), pp. 16.

David Macaulay, *The Way Things Work* (Boston, MA: Houghton Mifflin, 1988) pp. 230-237.

Guitar Tuner and

Loudspeaker Response

Questions

Q1. How are musical pitches related to frequency?

_____

Q2. Given that "A" is 440 Hz, calculate the frequency of "C" which is three half-steps higher.

_____

Q3. What is the relationship between a cent and the constant M?

_____

Q4. What would you expect the frequency response of a great loudspeaker to look like?

_____

Conclusions (What did you learn from this experiment?)

_____

_____

_____

Data and Observations

(Graph: Pitch (E, B, G, D, A, E) vs Frequency (Hz) 100–1000 on horizontal axis; Sound Pressure Level (dB_SPL) 60–110 on right vertical axis)

Signature: _____ Date: ___/___/___ Witness: _____ Date: ___/___/___

Prelab Questions: Guitar Tuner and Loudspeaker Response

(Bring sheet with questions answered to your lab session)

Print your name (Last, First): _____

Q1. Do you play a musical instrument? If your answer is "no", answer pre-lab question #2 below. If your answer is "yes", what instrument(s)? Could you use a "guitar tuner" in tuning your instrument(s)? Explain.

Q2. How do musicians "tune" an instrument such as a guitar? What about instruments such as the kettledrum and the trumpet — do musicians actually "tune" them? (Ask friends who play musical instruments, or look it up if you don't know.)

Lissajous 'Scope Patterns

Lab 6

Instructional Objectives (At the end of this lab you should be able to:)

I1. Measure the relative frequencies of two time-varying waveforms with a scope by forming and interpreting Lissajous (pronounced "lissa-jew") patterns.

I2. Define the ratios of audio frequencies that are given in terms of octaves, major and minor thirds, fourths, fifths, and cents.

Equipment

Analog oscilloscope that accepts both vertical and horizontal inputs; two audio-frequency signal generators; two loudspeakers; electronic keyboard; connecting wires and cables.

Description and Background

A standard method of examining time-varying voltage waveforms whose frequencies are simply related is the formation on a scope of a so-called Lissajous pattern. This pattern is formed when the electron beam of the scope's cathode-ray display tube is deflected in the vertical direction by one signal and in the horizontal direction by the other signal. If the two signals have identical frequencies and equal amplitudes, the Lissajous pattern will be a simple circle. If the frequencies are identical but the amplitudes differ, the pattern will be an ellipse; the "tilt" of the ellipse will depend on the relative phases of the two signals. If the two signals have unequal frequencies that are simply related — such as having frequencies that are simple integral multiples of each other — more complex patterns will be formed, such as "bow ties" from which the frequency ratios may readily be determined.

In this experiment, you will hear the signals as well as display them on a scope. This, together with the use of an inexpensive electronic keyboard, permits the study of some elements of musical intervals.

COMMENTS ON MUSICAL INTERVALS:

(Be sure you've read the **Description and Background** of Lab 5, "Guitar Tuner and Loudspeaker Response".)

The notes of the conventional Western musical scale are denoted by the letters A through G and by "modifiers" known as flats (represented by the symbol ♭) and sharps (represented by the symbol ♯). On a piano or synthesizer keyboard the unmodified notes are the white keys, and the modified notes are the shorter black keys. The notes used are: A, A♯ or B♭, B, C, C♯ or D♭, D, D♯ or E♭, E, F, F♯ or G♭, G, and G♯ or A♭. The white keys played in sequence from one A to the next form a "minor" scale. If played from one C to the next they form a "major" scale. Because of the importance of major scales, C is often used as a starting point for orientation. It is located immediately to the right of a white note and immediately to the left of a cluster of two black notes separated by a white note and followed by two adjacent white keys.

Procedures

P1. Connect one signal source to a loudspeaker and to the vertical 'scope input. Connect the other signal source to the other loudspeaker and to the horizontal 'scope input. Set the first signal source so that it produces a sound whose frequency is identical to that of the lowest C note on the tone synthesizer. The whistle tone is the closest to being a pure sinusoid. Without looking at the scope display, adjust the other source until the sound it produces has the same frequency. Now observe and interpret the Lissajous pattern displayed.

P2. Leave the first signal source set as before. Tune the second so that it matches the frequencies of other notes related to the C note by standard musical intervals, and observe and interpret the Lissajous patterns formed in terms of the ratios of the frequencies. Two notes that are an octave apart have frequencies whose ratio is 2:1. Each musical interval, thirds, fifths, etc., have an ideal integral ratio. Measure and identify the ratio for each interval.

P3. Look at the patterns in the **Data and Observations** section. For each of those patterns, identify the ratio of the frequencies and the corresponding musical interval. (Do not use "hardware" — the frequency generators — to do this. Convince yourself that the ratio of the frequency of the vertical axis source to that of the horizontal axis source equals the ratio of the number of crossings of the horizontal axis to those of the vertical axis.)

References

A. H. Benade, *Fundamentals of Musical Acoustics*, 2nd Revised Ed. (New York: Dover Publications, 1990).

P. Horowitz and W. Hill, *The Art of Electronics*, 2nd ed. (Cambridge, England: Cambridge University Press, 1989), Appendix A, pp. 1045-49.

David Macaulay, *The Way Things Work* (Boston, MA: Houghton Mifflin, 1988), pp. 230-237.

Lissajous 'Scope Patterns

Questions

Q1. Which intervals sound most pleasant to you?_____

Q2. How large are the errors introduced by the "even tempered scale," where each half step is exactly $^{12}\!\sqrt{2}:1$, when compared with the ratios of integers found in the lab? Express your answers in cents.

_____

Conclusions (What did you learn from this experiment?)

_____

_____

Data and Observations

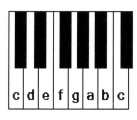

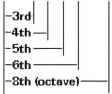

$M = 1.059 = \sqrt[12]{2}$

$M^4 = 1.26 \approx \dfrac{5}{4} = 1.25 = \textbf{3rd}$

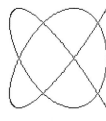

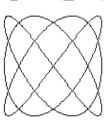

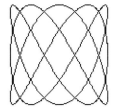

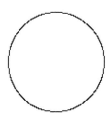

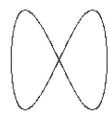

Signature: _____ Date: ___/___/___ Witness: _____ Date: ___/___/___

Prelab Questions: Lissajous 'Scope Patterns

(Bring sheet with questions answered to your lab session)

Print your name (Last, First): _____

Q1.   Sketch the Lissajous pattern you expect for each of the following pairs of signals:

a. Vertical input: $A \times \sin(2\pi f_1 t)$; horizontal input: $A \times \cos(2\pi f_1 t)$.

b. Vertical input: $A \times \sin(2\pi f_1 t)$; horizontal input: $A \times \sin(2\pi [f_1 \times 2]t)$.

Q2.   What is the frequency ratio of two tones that are two octaves apart?

Touch-Tone® Telephone

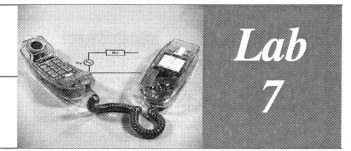

Lab

7

Instructional Objectives (At the end of this lab you should be able to:)

I1. Describe how tone signaling works.

I2. Build a Thévenin equivalent circuit.

Equipment

Telephone; audio signal generator; digital volt meter (DVM); analog (x-y) oscilloscope; 0-35VDC power supply; resistors; various phone jacks and adapters (optional: frequency counter).

Description And Background

Touch-Tone® telephones have been in use since the late 1950s. They offer convenience not only in ease of dialing, but in operating equipment remotely, such as voice mail systems.

These phones signal by generating a pair of sinusoidal tones each time a key is pressed.

Procedures

P1. Measure the telephone line coming into the lab. Determine open-circuit voltage and short-circuit current with the DVM. Note: The measurement of short-circuit current is allowable because it will not harm the telephone company's circuitry. DO NOT try this with the electric power system.

P2. Build a Thévenin equivalent circuit to replace the phone company, so that you can run the telephone on the bench. Listen for touch tone signals. Also sketch the Norton equivalent circuit. The phone will not generate tones if connected with reverse polarity (there will be no dial tone).

P3. Configure the oscilloscope to generate Lissajous patterns using the signal generator as the x-axis signal and the telephone signal as the y-axis signal. Tune the generator to form stretched ring patterns that indicate matching of one of the two tones. Fill in all 24 frequencies on the phone pad indicated in Table 1 on page 57.

References

P. Horowitz and W. Hill, *The Art of Electronics*, 2nd ed. (Cambridge, England: Cambridge University Press, 1989), pp. 727 919; 933-37.

David Macaulay, *The Way Things Work* (Boston, MA: Houghton Mifflin, 1988), pp 252-253.

Touch-Tone® Telephone

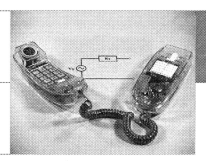

Questions

Q1. What is the pattern of the frequencies that are used?

_____

Q2. What is the tolerance (acceptable error) on these frequencies?

_____

Q3. Why are two tones used instead of just one?

_____

Q4. Why can't we measure the frequencies with a counter?

_____

Q5. What trick, which works on old AT&T phones, will enable us to measure these with a counter or an oscilloscope?

_____

Conclusions (What did you learn from this experiment?)

_____

_____

Data and Observations

Table 1: Touch-Tone Frequencies

1	2	3
___ ___	___ ___	___ ___
4	5	6
___ ___	___ ___	___ ___
7	8	9
___ ___	___ ___	___ ___
*	0	#
___ ___	___ ___	___ ___

Signature: _____ Date: ___/___/___ Witness: _____ Date: ___/___/___

Prelab Questions: Touch-Tone® Telephone

(Bring sheet with questions answered to your lab session)

Print your name (Last, First): _____

Q1.   Why do you think that <u>two</u> tones are used to represent each number?

Curve Tracer

Lab

8

Instructional Objectives (At the end of this lab you should be able to:)

I1. Use a curve-tracing oscilloscope (curve tracer) to display current-voltage (*I-V*) characteristics of important electronic components such as resistors, diodes, and transistors.

I2. Measure the turn-on voltages of pn silicon and light-emitting diodes (LEDs).

I3. Measure the amplification factor of a bipolar junction transistor (BJT) and a field-effect transistor (FET).

I4. Identify from its curve tracer characteristics the nature of an unknown simple electronic component

Equipment

Curve tracer (such as Tektronix Model 576); diodes (pn, zener, visible LED, infrared LED); BJTs; FETs; pn silicon solar cell; unknown components; infrared sensor card; multimeter (ohmmeter, and diode tester if available).

Description and Background

In order to measure the *I-V* characteristics of electronic components one often uses the curve tracer. This instrument applies a time-varying voltage to a device under test (DUT); the curve tracer then displays the current that flows (vertical axis) versus the applied voltage (horizontal axis). Thus one can readily obtain plots of the *I-V* characteristics of one-port (two-terminal) components such as resistors and diodes.

For three-terminal devices, such as transistors, the curve tracer employs an additional feature: it provides an additional voltage (or current) that is stepped repetitively through a set of prescribed values; this stepped voltage is applied to the third terminal of the DUT, and causes a family of *I-V* characteristics to be generated and displayed. For example, the stepped voltage can be applied to the gate of an FET as the curve tracer displays the current flowing through the FET channel as a function of the applied source-drain voltage.

The sketch at the end of these notes shows schematically how the curve tracer operates. In the simplest (and least expensive) curve tracers, a very simple, low-frequency oscilloscope is used. The time-varying voltage applied to cause current flow in the DUT is simply derived from the 60 Hz powerline voltage. Thus, the applied voltage sweeps from zero to its maximum value 60 times per second. This produces both current flow in the DUT and deflection of the scope beam in the horizontal direction. Many of the curve tracer knobs control the sensitivity of the horizontal and vertical sweeps. Additional controls on the instrument can be used to limit the voltage that can be applied to the DUT (to protect it from being burned out), adjust the polarity of the applied voltage (positive or negative), and adjust the value of a series resistance inserted between the DUT and the swept voltage source. For three-terminal DUTs, the number and magnitudes of the stepped control voltage can be adjusted.

Procedures

P1. <u>Familiarization.</u> Take a few minutes to look over the many controls on the curve tracer. Many will be familiar from your earlier work with the oscilloscope, such as the beam controls (intensity, focus) and the vertical and horizontal sensitivities. Additional switches and knobs control the characteristics of the voltages applied to two- and three-terminal devices. An important switch to note is the three-position toggle switch at the lower right of the instrument (it is marked "LEFT", "OFF", "RIGHT"); when in its center, vertical position, no voltage is applied to any DUT plugged into the test sockets to the left. When putting a different DUT in a test socket you should flip this switch to its vertical off position.

For all the tests in this experiment set the maximum voltage to its lowest value: turn the "double" knob marked "Max. Peak Volts" and "Series Resistors", to its minimum (zero, fully counterclockwise) position and return it to this position when you complete each measurement. Also turn the knob labelled "Variable Collector Supply (Percent of Max. Peak Volts)" to zero after each measurement. This will help reduce the number of devices that become accidentally overheated! YOU WILL BE GIVEN ONLY <u>ONE</u> OF EACH

OF THE MOST SENSITIVE COMPONENTS TO TEST, SO BE CERTAIN THAT YOU DON'T APPLY EXCESSIVE VOLTAGES TO THEM.

P2. Resistors. Set the horizontal sensitivity to about two volts full scale (0.2 volts/div.), and the vertical sensitivity to no more than 10 mA full scale (1 mA/div.). Set the "Series Resistor" control to its minimum value, which is typically 0.3 ohms. (You can set the series resistance value without changing the collector voltage value by pulling the knob outward when changing its position.)

Flip the toggle switch to its vertical OFF position. With no DUT in the test socket, flip the toggle to the left and observe the *I-V* display. Is it what you expected?

Now flip to the OFF position again and connect a resistor between the curve tracer's emitter and collector terminals. To do this, either (a) install a resistor in the adapter socket labelled "BCE" (it's made for BJTs), putting the resistor leads in the C and the E "holes"; or (b) connect a resistor in your Science Fair 130-in-one Electronic Project Lab to the C and E points.
Flip the toggle switch to the ON position, and observe and record what the display indicates. From the display calculate the value of resistance. Record this value and compare it with the values you get from an ohmmeter measurement and from the color code indication on the resistor. Do this with at least two resistors whose resistances differ substantially.

P3. Diodes.   Use the same horizontal and vertical sensitivity settings you used in P2. Set the series resistance to 3KΩ or 14KΩ, as described above. Find the turn-on voltage in the forward direction. Also observe the *I-V* characteristic in the reverse direction, applying up to about 15 volts to the diode to see whether it exhibits significant current flow in the reverse direction.

P3a. Silicon pn diode. Flip the toggle switch to OFF and plug a silicon pn diode into the BJT socket between connection holes C and E, putting the negative lead in the E hole. (The negative lead protrudes from the end of the diode that has a bar or continuous band around it, and on some diodes the negative lead is the shorter lead.) If the "Polarity" switch is in the "+(NPN)" position, the voltage V_{CE} will be positive and a current flowing into C will be plotted as being positive. In other words, the screen display will be of the first quadrant of *I-V* space. With the polarity switch in the "-(PNP)" position, you get a plot of the third quadrant.

Never exceeding either 20mA or 15 volts, gradually increase the AC voltage applied to the diode to display the forward and reverse *I-V* characteristics. Determine and record the turn-on voltage, as well as any interesting reverse characteristics (also record the diode type if you can read it on the diode itself).
P3b. Silicon zener diode. If one is available, repeat P3a with a zener diode. A zener diode conducts in the forward direction (easy current flow) much as a conventional silicon pn diode, but it also conducts at some well-defined voltage when driven in the reverse direction by a voltage of opposite polarity.
P3c. Visible LED. Repeat with an LED that emits visible light. Note: These are made from gallium arsenide (GaAs), not silicon, so the diode will have a different turn-on voltage. Note how the voltage for light emission relates to the voltage at which the current rises from its very low value. Record your data, including the color of the light emitted.
P3d. Infrared LED. Repeat with an infrared LED, using the infrared sensor card to determine when this LED is emitting.

P4. BJT.   The transistor is a three-terminal device; to display its characteristics fully you will also use the stepped current source provided by the curve tracer.

Put the transistor into the socket provided, being sure that the leads are correctly positioned (base lead in the B hole, etc.). Adjust the polarity switch to npn or pnp to match your transistor type. Set the series resistor at its minimum value, operate with the "EMITTER GROUNDED" and with the "STEP GEN" on. Set "STEP GEN" to 10μA/step. Set "NUMBER OF STEPS" to 8. From the characteristics displayed and the sensitivities for the display (marked on the knobs and repeated in lights at the right side of the display), determine the actual amplification factor for the transistor you test. (Recall that the amplification factor, β, is defined from the relation between collector and base currents, $I_C = \beta I_B$. The value in lights to the right of the screen is not necessarily the value for your transistor.)

P5. <u>FET.</u> **BE CAREFUL NOT TO CONTACT ANY OF THE FETs ELEMENTS BY TOUCHING ITS OWN LEADS OR THE PLUGS ON THE SPECIAL FET HOLDER THAT PLUGS INTO THE CURVE TRACER. STATIC CHARGE OR INDUCED AC POTENTIALS ON YOUR BODY COULD DESTROY THE INSULATING PROPERTIES OF THE FET GATE INSULATOR.** This is why these devices are often shipped in a conductive plastic bag or pushed into a piece of black electrically conductive foam.

Carefully plug the FET holder into the curve tracer. Set up the controls to display I_C vs. V_{CE} with V_{GS} (voltage) as the stepped parameter instead of current as you did for the BJT. Determine the amplification factor of the FET you were given.

P6. <u>Unknown.</u> Obtain from your instructor an unknown electronic component and carefully measure its forward and reverse characteristics. Try to determine both qualitatively and quantitatively the nature of the component.

P7. <u>Load line.</u> (Optional) With either the BJT or the FET in the curve tracer you see an effect that is closely related to the important concept of the load line. As shown in the diagram of curve tracer operation (Figure 8.1), the horizontal sweep indicates the voltage (V) applied to the DUT. For any given supply voltage (i.e., for a given setting of the "VARIABLE COLLECTOR SUPPLY" knob), V depends on the current (I) flowing in the DUT and upon the value (R_{SER}) of the series resistance used; the relationship is just

$$V = V_S - I R_{SER}$$

Thus, your transistor I-V characteristics terminate at different values of V, depending upon how much current flows as the base or gate voltage is stepped. The line formed by connecting those terminating values is the load line for the voltage supply consisting of the voltage source V_S in series with an internal resistance R_{SER}. From the displays for one of your transistors, verify quantitatively that the "phantom" load line displayed is what you would predict from the settings you have used.

References

P. Horowitz and W. Hill, *The Art of Electronics*, 2nd ed. (Cambridge, England: Cambridge University Press, 1989), Appendix A, pp. 1045-49; also see pp. 13-15, 44, and 932-3.

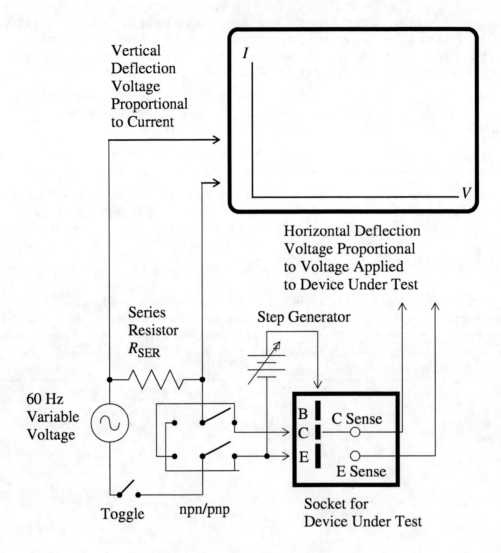

Vertical
Deflection
Voltage
Proportional
to Current

I

V

Horizontal Deflection
Voltage Proportional
to Voltage Applied
to Device Under Test

Series
Resistor
R_{SER}

Step Generator

60 Hz
Variable
Voltage

B
C
E

C Sense

E Sense

Toggle   npn/pnp

Socket for
Device Under Test

Figure 8.1   Simplified Diagram of the Curve Tracer

Curve Tracer

Questions

Q1. How might you use a curve tracer if you were doing transistor device research for the Intel Corporation?

_____

Q2. How might you use a curve tracer if you were designing a consumer stereo?

_____

Conclusions (What did you learn from this experiment?)

_____

_____

_____

Data and Observations

Signature: _____ Date: ___/___/___ Witness: _____ Date: ___/___/___

Pre-lab Questions: Curve Tracer

(Bring sheet with questions answered to your lab session)

Print your name (Last, First): _____ Date: __/__/__

Q1. The curve-tracer plots on its horizontal axis the voltage applied to the device under test (DUT) and on its vertical axis the current that then flows through the DUT. Sketch below the curve-tracer displays you expect (assuming that the horizontal sensitivity is 2 V/division and the vertical sensitivity is 1 mA/division) for each of the following:

DUT #1: 2000-ohm resistor

DUT #2: 8000-ohm resistor

DUT #3: Conventional silicon pn diode

Curve-tracer display #1   Curve-tracer display #2   Curve-tracer display #3

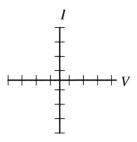
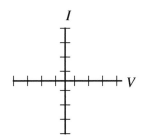
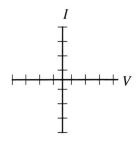

Astable Multivibrator

Instructional Objectives (At the end of this lab you should be able to:)

I1. Understand how a multivibrator circuit oscillates.

I2. Figure out the formulas for multivibrator operation.

(Don't attempt this lab until you understand how transistors operate.)

Equipment

Science Fair 130-in-one Electronic Project Lab; oscilloscope.

Description And Background

The Glossary defines a multivibrator as "a circuit that uses feedback to provide a repetitively changing output". In the *Science Fair 130-in-one Electronic Project Lab*, multivibrator circuits are used to produce buzzing sounds and other audio effects. In essence, each multivibrator circuit contains a pair of transistors connected so that the output of one (call it Q_1) is coupled to the input of the second transistor (Q_2) (see Figure 9.1). In addition, the output of Q_2 is coupled to the input of Q_1. Because of this sort of cross connection, the transistors never reach a stable condition in which one is conducting and the other is not conducting. The action is like that of a teeter-totter that never stops.

This lab provides a good opportunity to explore bipolar transistors operating in a digital mode. You will also see capacitors used in two ways — controlling a time constant and coupling time-varying voltages.

To analyze the circuit assume that one transistor is on and the other is off. Assume a V_{beON} of -0.7 volts. When the transistor is off, V_{be} can be anything above -0.7 volts *including positive*.

Procedures

P1. Build an astable multivibrator circuit with the *Science Fair 130-in-one Electronic Project Lab* (see the "Horror Movie Sound Effect"). Draw the schematic of just the astable multivibrator.

P2. Observe its operation with two channels of a oscilloscope. Sketch the waveforms at the transistor terminals (use the emitters as "ground").

P3. Figure out how it works. Give a step-by-step description of its operation emphasizing cause and effect (assume any initial conditions, and state them).

References

Science Fair 130-in-one Electronic Project Lab (Fort Worth, TX: RadioShack [Tandy Corp], 1989), Exp. #12, "Horror Movie Sound Effects"; Exp. #17, "Resistors in Series and Parallel"; Exp. #22, "Flip-Flop Multivibrator with LED Display"; Exp. 32, "Transistor Flip-Flop Circuit".

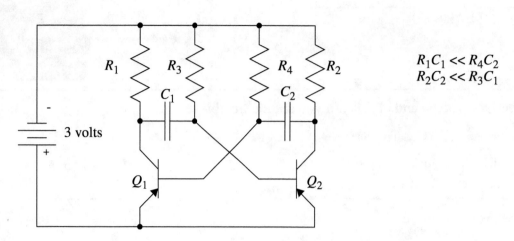

$$R_1C_1 \ll R_4C_2$$
$$R_2C_2 \ll R_3C_1$$

Figure 9.1   Basic astable multivibrator circuit

Astable Multivibrator

Questions

Q1. Which resistors control the time constants?

_____

Q2. Determine what fraction of the *RC* time constant is actually used. Write out a formula for the operating frequents in terms of the components values.

_____

Conclusions (What did you learn from this experiment?)

_____
_____
_____

Data and Observations

Signature: _____ Date: ___/___/___ Witness: _____ Date: ___/___/___

Prelab Questions: Astable Multivibrator

(Bring sheet with questions answered to your lab session)

Print your name (Last, First): _____

Q1. What appears to be the minimum number of transistors that one needs to make an astable multivibrator (you can check in the manual for your Science Fair 130-in-one Electronic Project Lab)?

Q2. What does the name "transistor" mean? (Note that the answer is in your Text.)

Infrared Door Alarm

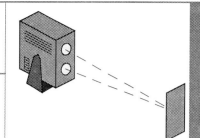

Instructional Objectives (At the end of this lab you should be able to:)

I1. Be able to design and astable multivibrator circuit.

I2. Understand operation of this class of security devices.

Equipment

Door alarm unit; corner retro-reflector; Science Fair 130-in-one Electronic Project Lab; oscilloscope; IR phototransistor; IR LED.

Description And Background

Security systems based on the use of an infrared (IR) beam that an intruder unknowingly interrupts have the obvious advantage that the unaided human eye cannot detect this radiation (see discussion about IR in the lab "Infrared Remote Control"). Fortunately, it is easy to generate and detect IR radiation, with an IR light-emitting diode and a phototransistor, respectively. This helps to make the use of IR radiation in security systems economically viable.

If you're curious about IR, you might be interested in the following information. (1) In addition to phototransistors, some electronic camera chips used in camcorders are sensitive to IR radiation. (2) A company has recently developed an inexpensive IR camera chip that produces a visible image at television scan rates from the IR radiation emitted by natural objects such as trees and buildings. It is planned to use this chip as the basis of a vehicle driving aid that permits one to navigate at night or in fog. (3) Infrared radiation can penetrate a distance of about a millimeter in a semiconductor such as silicon, even though visible light cannot. Therefore, when making some silicon electronic devices by photolithographic means, one can align a pattern on top of the silicon wafer with one on the bottom by shining IR radiation right through the wafer.

Procedures

P1. Build a small IR sensing circuit, like the one you used in the IR remote control lab.

P2. Measure the output of the door alarm.

P3. Build an astable multivibrator to drive an IR LED at the measured frequency.

P4. Explain how the circuit works.

References

Lab 9, "Astable Multivibrator" on page 69
Lab 3, "Infrared Remote Control" on page 31

Infrared Door Alarm

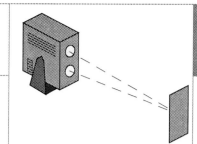

Questions

Q1. Measure the distance that your circuit will override the alarm.

_____

Q2. How does the special "corner reflector" work?

_____

Conclusions (What did you learn from this experiment?)

_____

_____

_____

Data and Observations

Signature: _____ Date: ___/___/___ Witness: _____ Date: ___/___/___

Prelab Questions: Infrared Door Alarm

(Bring sheet with questions answered to your lab session)

Print your name (Last, First): _____

Q1. List three reasons for using infrared instead of visible light or radio in such a door alarm.

_____

_____

_____

_____

_____

_____

Television

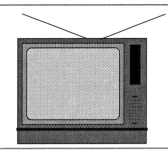

Instructional Objectives (At the end of this lab you should be able to:)

I1. Describe how television works, including the rate at which individual fields are presented, interlacing, the use of three basic colors to produce all displayed colors, and to define the terms luminance, chrominance, synchronization signals, hue, and saturation.

Equipment

Video cassette player; oscilloscope (with delayed sweep and TV triggering); VHS tape of test images; television receiver; connecting cables.

Description And Background

Television displays are ubiquitous as are video cassette players. In this lab we explore the encoding of a picture in the NTSC signal.

Bear in mind that this standard was first developed for "Black and White" (actually greyscale) television. This should lead you to understand that the parts of the signal used to encode this type of picture should be simpler than the parts that encode the color information. The basic synchronization information that marks the edge of the picture is unchanged between b&w and color.

The color-space that is implemented by this standard uses three values to quantify a color. These are: hue, a number expressed in angular coordinates (0-360 degrees) that varies from red through the shade of the spectrum to violet and then wraps on around through shades of magenta back to red; saturation, a measure of how strong or pure the color is; and lightness, which is measure of how light or dark the color is.

Procedures

P1. Connect an oscilloscope that features TV triggering (typically having TVL and TVF trigger modes) to the video output of the VCR. Don't try to use the antenna or RF output. Play the video tape to generate a predictable and repeatable signal. Set the 'scope's main time base to 2ms/div. Sketch the waveform.

P2. Figure out the correspondence between position in the signal and position on the screen.

P3. Use the oscilloscope's delayed time base on 10µs/div. to observe finer details. Again figure out position correspondence. Identify on your sketches: vertical sync pulses, horizontal sync pulses, black level, white level, and gray signal.

P4. Use the color parts of the test tape to measure color signals. Identify the color burst and saturated color signals.

References

Gordon McComb, *Troubleshooting & Repairing VCRs*, 2nd ed., (TAB Books, 1991), pp. 21-27.

David Macaulay, *The Way Things Work*, (Houghton Mifflin, 1988), "Television Camera," pp258-9; "Video Recorder," pp. 260-1; "Television Set", pp. 262-3.

Dick White and Roger Doering, *Electrical Engineering Uncovered*, (Prentice Hall, 1997) See "How Many Words Is a Picture Really Worth?" on page 27.

Robert Schetgen, ed., *Handbook for Radio Amateurs*, 70th ed., (American Radio Relay League, 1993), pp. 20-2.

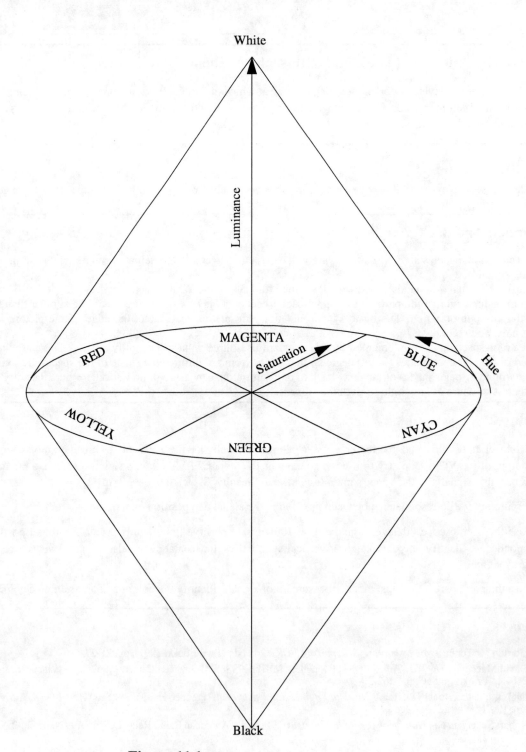

Figure 11.1 Color space co-ordinates

You may which to further annotate these figures to turn in with your lab report.

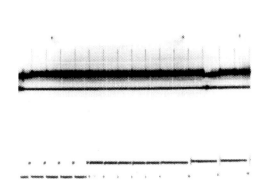

Figure 11.2 Top — entire video field of white circle and grid on black background. Bottom — detail of vertical sync.

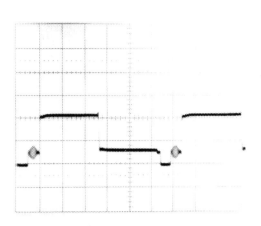

Figure 11.3 Detail of horizontal line showing horizontal sync, color burst, white level and black levels.

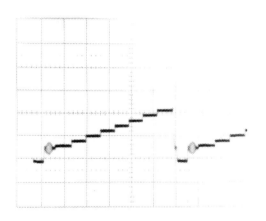

Figure 11.4 Horizontal line showing eight gray levels progressing from black to white.

Figure 11.5 Horizontal line showing color stripes: white, yellow, cyan, magenta, green, red, blue and black.

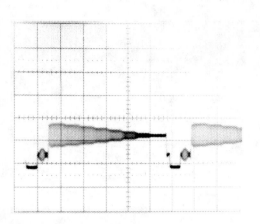

Figure 11.6 Horizontal line showing a single color hue progressing from fully saturated to gray.

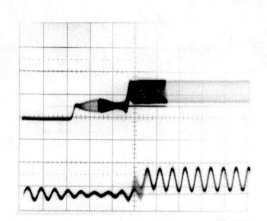

Figure 11.7 Top — horizontal sync, color burst and solid color field. Bottom — enlargement of color burst and solid color.

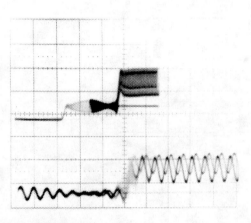

Figure 11.8 Similar to Figure 11.7 except showing two different hues overlaid to show phase difference. The two colors shown also have different luminance and saturation levels.

Television

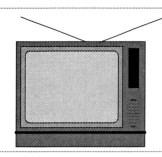

Questions

Q1. What are the frame, field, and line rates?

_____

Q2. How are intensity, saturation and hue encoded?

_____

_____

Q3. What does NTSC stand for (check in _Electrical Engineering Uncovered_)?

_____

Conclusions (What did you learn from this experiment?)

_____

_____

_____

Data and Observations

Signature: _____ Date: ___/___/___ Witness: _____ Date: ___/___/___

Prelab Questions: Television

(Bring sheet with questions answered to your lab session)

Print your name (Last, First): _____

Q1. When was color television first demonstrated or used? (Note: the answer is in *Electrical Engineering Uncovered*.)

Q2. What does "saturation" mean in connection with color reproduction?

VCR

VCR
Insert this side into recorder ↓ Do not touch the tape inside
Sleeper

Lab

12

Instructional Objectives (At the end of this lab you should be able to:)

I1. Describe the operation of a helical head recorder.

Equipment

VCR; any tape; infrared LED; infrared phototransistor; infrared sensor card; Science Fair 130-in-one Electronic Project Lab; clipleads; tape measure; stopwatch.

Description And Background

This appliance has become quite commonplace in the home. The technology that makes it possible to record high bandwidth video signals is the rotating record/playback head. This motion achieves a relative speed between head and tape that is extremely high. Audio and control tracks are positioned along the edges of the tape, and are recorded and played back using separate stationary heads.

Information recorded on a tape is represented by tiny, differently magnetized grains of iron oxide or other metals on the tape. A magnetic nickel-iron alloy is used to form a ring-shaped magnetic tape write/read head. The head is a ring of permalloy having a tiny gap cut radially in it for positioning near the tape. A copper coil wound around the ring produces a magnetic field in the gap during the writing process, when the coil is supplied with a current. During reading, magnetized regions of tape passing by the gap produce a magnetic flux in the ring, generating a voltage at the terminals of the coil. In a stereo VCR audio head, two such cores are embedded in a non-magnetic ceramic material to form a smooth surface for the tape to slide on. The moving video head will have two or four heads that are used sequentially.

Procedures and Questions

P1. Remove the cover of the VCR. Watch the mechanism load and unload tapes. Draw the tape path and identify the following: pinch roller, capstan, erase head, audio/control head, video head, and tape guides.

P2. Construct a pair of simple circuits: one to emit infrared light and one to detect it. Connect the phototransistor and LED using clip-leads so that you can move them around. Using these and an oscilloscope, bounce light off of the top of the rotating head where the holes pass by. You will need to position the devices very close to the head (< 1 cm.) The holes will not reflect the light, and you'll get a waveform on the oscilloscope. Measure the speed of rotation of the moving head. Measure the tape speed using the white dot on the pinch roller.

P3. Calculate the relative speed between tape and head.

P4. Calculate the angle of the tracks on the tape. Sketch the tape format (include audio and control tracks). Calculate the track-to-track distance (orthogonal to the track).

References

David Macaulay, *The Way Things Work* (Boston, MA: Houghton Mifflin, 1988), "Video Recorder," pp. 260-1

Gordon McComb, *Troubleshooting & Repairing VCRs*, 2nd ed. (Blue Ridge Summit, PA: TAB Books, 1991), pp. 14-20; 34-5.

VCR

VCR

Insert this side into recorder ▲ Do not touch the tape inside

Sleeper

Questions

Q1. Why does the head rotate at the observed speed?

_____

Conclusions (What did you learn from this experiment?)

_____
_____
_____

Data and Observations

Signature: _____ Date: ___/___/___ Witness: _____ Date: ___/___/___

Q1. How are video and audio signals recorded on a VCR cassette tape?

Q2. Why does the video read/write head spin rapidly even though the tape moves through the machine relatively slowly? (You might get a clue from reading "How Many Words Is a Picture Really Worth?" in *Electrical Engineering Uncovered*.)

RC Filters

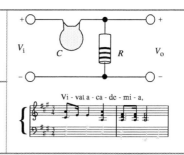

V_i C R V_o

Vi - vat a - ca - de - mi - a,

Instructional Objectives (At the end of this lab you should be able to:)

I1. Identify highpass and lowpass RC filters and their characteristics.

Equipment:

Science Fair 130-in-one Electronic Project Lab; LCR meter; signal generator; oscilloscope.

Description and Background

The simplest electrical *filters* are two-port devices that, when supplied with an input signal, produce an output signal whose amplitude depends upon both the amplitude and the frequency of the input signal.

As an example of filtering, suppose that a microphone is actuated by human speech and that its output is fed to a lowpass filter operating in the audio range. The output of this lowpass filter would contain only the low-frequency components of the speech, and the high-frequency components would be lost. Conversely, a highpass filter would let the high-frequency components pass through. Examples of the transmission characteristics of lowpass and highpass filters are shown in the "Simulink" lab section, "Transfer Functions for Basic R-L and R-C Filter Circuits" on page 168.

Your home music system may use a so-called crossover filter to separate the low-frequency tones and the high-frequency tones for the woofer and the tweeter (see the Glossary in *Electrical Engineering Uncovered* for definitions of "woofer" and "tweeter"). The crossover filter contains one lowpass and one highpass filter. The inputs of both filters are connected together, but the output of the lowpass filter goes to the woofer, and the output of the highpass filter goes to the tweeter. (See the lab "Loudspeaker Crossover" on page 119.) One can also make a bandpass filter that only lets through components whose frequencies lie between the filter's lower and upper cutoff frequencies.

In this experiment you will explore the characteristics of lowpass and highpass filters made using one resistor and one capacitor. In thinking about the operation of these RC filters, you might consider how the corresponding water models of these circuits would operate. For example, if the output is the capacitor voltage (or capacitor pressure difference, in the water model) as in Figure 21.10 on page 150 in *Electrical Engineering Uncovered*, the time constant for charging (filling) the capacitor is the product RC. At low enough input signal frequencies, for which the period of the wave is much less than the time constant RC, there is time to charge (fill) the capacitor and obtain an output. At very high frequencies, where the period of the wave is much smaller than the time constant, the capacitor charging can't follow perfectly the input voltage (pressure) variations and so the output is very small. Thus, this is a lowpass RC filter (series R, C in parallel with the output terminals).

When recording your data in this lab, be sure to note the values of R and C that you use, since their product determines the time constant and hence the breakpoint where the filter changes between transmitting well and transmitting poorly. Identify which of your filters is lowpass and which is highpass. Find the break frequency for each circuit (this is $2\pi f_{3dB} = \omega_{3dB}$). Also, use the proper kind of graph paper so that the output plotted versus frequency consists of straight lines below and above the cutoff frequency.

WATER MODELS

Since electricity can't usually be seen, for purposes of visualization it is convenient to consider water models for electrical quantities (see *Electrical Engineering Uncovered* for more discussion of these water models). Electric current flowing in a wire is analogous to the flow of water in a pipe. The voltage across a circuit element — the potential difference between the two ends of the element in a circuit — is analogous to a difference of water pressure between two points in a network of plumbing. The pressure difference might be produced by a pump driven by a constant-torque motor (Figure 13.1). The water models for a resistor, capacitor and inductor are shown in Figure 13.2.

Figure 13.3 shows a complete water-model RC circuit. The voltage source is powered by a constant torque motor, as described earlier. The resistor is a pipe containing a constriction, and the capacitor is the chamber fitted with a piston and a spring. Consider the charging" of this capacitor by the voltage source. Let us suppose that the piston of the capacitor is initially in its equilibrium position in the middle of the chamber. In other words, when we start there is no pressure difference (no voltage) across the capacitor. Now we turn on the constant torque motor and start to fill the capacitor, pushing the piston to the right and establishing a pressure difference between the two ends of the capacitor. Let's think about how the time to drive the piston to a given position, say halfway to the right of the midpoint of the chamber, would depend on the values of the resistor and the capacitor

First, if the resistance is high (if the pipe is severely constricted), it will take a long time to force a given amount of water through it and into the capacitor. Thus, we'd expect that the time required for charging the capacitor would rise as the value of the resistance rises. If instead the capacitance were to increase, it would take more water to push the piston to the three-quarter position; since getting more water in would require more time, the time required for charging should also increase as the capacitance increases. In fact, the time constant — the time for the capacitor to charge to about 63% of its maximum value — is simply equal to the product RC.

Procedures

P1. Connect a 10kΩ resistor and a (non-polarized) 0.1 µF capacitor in series with a signal generator, making sure that your oscilloscope ground and the signal generator ground are connected together. Set the signal generator to output a 1-volt peak sine wave. Measure and plot the amplitude of the voltage between the components versus frequency on log-log graph paper.

P2. Check out the effects of filtering on square and triangular waves.

P3. Reverse the order of the two components and repeat.

References

P. Horowitz and W. Hill, *The Art of Electronics*, 2nd ed. (Cambridge, England: Cambridge University Press, 1989), pp. 35-8.

Dick White and Roger Doering, *Electrical Engineering Uncovered* (Upper Saddle River, N.J.: Prentice Hall, 1997): "Logarithmic Unit for a Person's Pay: The Salarybel" on page 23 for an explanation of decibels; skim "Waveforms and Spectra" on page 235, and "Modeling Electrical Devices: The Transfer Function" on page 243, for further information about filters.

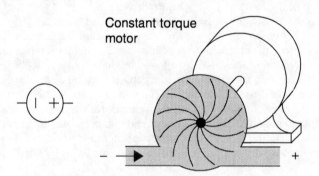

Figure 13.1 Water model for a steady voltage source (schematic symbol at left)

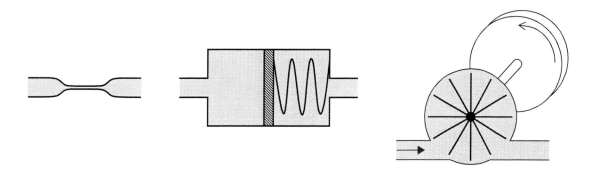

Figure 13.2   Water models and schematics symbols for a resistor (left), capacitor (center) and an inductor (right)

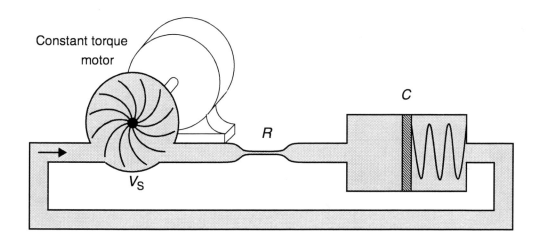

Figure 13.3   Water model of a complete circuit containing a steady voltage source, a resistor, and a capacitor

RC Filters

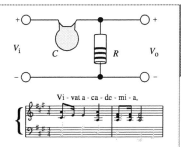

Questions

Q1. What formula determines ω_{3dB}, where the signal amplitude is 0.707 times the input amplitude?

_____

Conclusions (What did you learn from this experiment?)

_____

_____

_____

Data and Observations

Signature: _____ Date: ___/___/___ Witness: _____ Date: ___/___/___

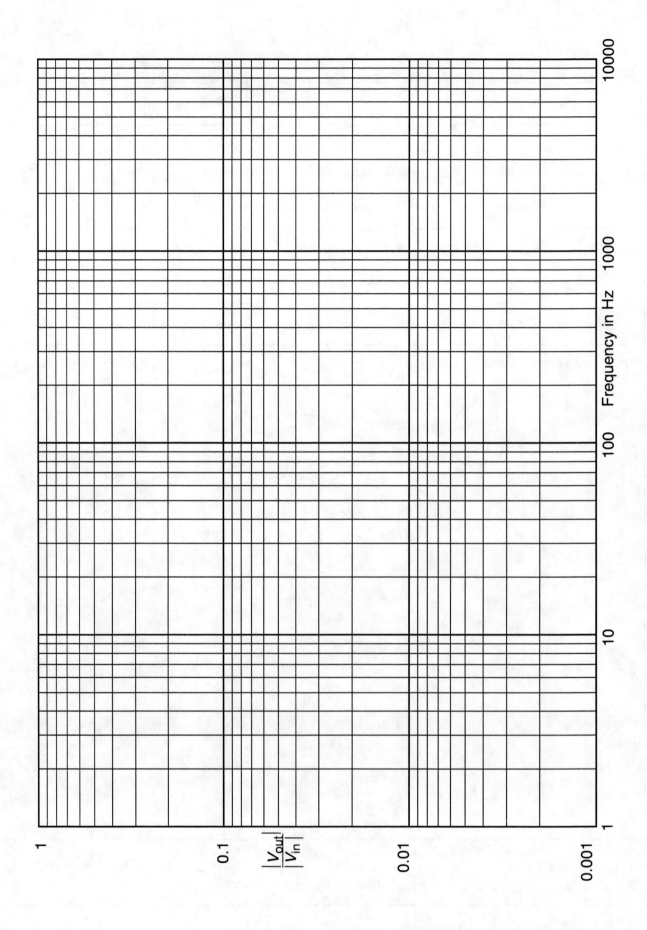

Prelab Questions: RC Filters

(Bring sheet with questions answered to your lab session)

Print your name (Last, First): _____

Q1. Why does the lab write-up specify that you should use an "non-polarized" capacitor?

Q2. If the ordinary frequency, in Hz, of a sinusoidal signal is 100 Hz, what is the corresponding angular frequency of that signal (both magnitude and units)?

Resonant Filters

Lab 14

Instructional Objectives (At the end of this lab you should be able to:)

I1. Calculate the resonant frequency of a filter containing one inductor and one capacitor, given the values of L and C.

I2. Measure the resonant frequency of a filter containing one inductor and one capacitor.

Equipment

Oscilloscope with a 50Ω input and X-Y mode; radio frequency sweep generator; four 50-ohm cables; BNC "T" connector; one BNC- to-cliplead connector; Science Fair 130-in-one Electronic Project Lab; two or more inductors; LCR meter.

Description and Background

The Card Key Lab has a good description of how resonant LC filters work. Besides card keys, there are many other applications for LC filters. For example, in AM radio transmitters and receivers there are "tank" circuits that resonate at one frequency better than others (bandpass filters). The capacitor or inductor is made variable so that this frequency can be easily changed to tune into different stations.

WATER MODELS

Since electricity can't usually be seen, for purposes of visualization it is convenient to consider <u>water models</u> for electrical quantities (see *Electrical Engineering Uncovered* for more discussion of these water models). Electric current flowing in a wire is analogous to the flow of water in a pipe. The voltage across a circuit element — the potential difference between the two ends of the element in a circuit — is analogous to a difference of water pressure between two points in a network of plumbing. The pressure difference might be produced by a pump driven by a constant-torque motor (Figure 14.1). The water models for a resistor, capacitor and inductor are shown in Figure 14.2.

We can visualize the behavior of a parallel-connected LC-resonant electrical circuit by considering its water model (Figure 14.3). The element on the left is the water model for the inductor. It consists of a turbine located in a close-fitting enclosure and connected to a massive flywheel whose inertia limits the speed at which the turbine can rotate. The element on the right is the water model for the capacitor. It is a close-fitting piston in a water-filled pipe fitted with a spring. To push more water in from the top (like adding charge to a "real" capacitor) requires more and more force — an increasing pressure difference between the two ends of the water capacitor — just as the voltage across a real capacitor increases as it is charged.

Suppose that we give the flywheel on the water inductor a little turn clockwise and then let go. The paddle wheel would start to turn clockwise, pushing water into the top of the capacitor. This will cause the spring in the capacitor to compress and produce an opposing pressure that will ultimately cause the water flow to reverse. The back flow through the inductor will push the piston up, ultimately stretching the spring so that the flow will finally reverse again. As there's no energy dissipating element in our plumbing, the flow will continue to oscillate forever.

The amount the piston moves or the turbine turns depends on the stiffness of the spring and the inertia of the flywheel. You can show that there is a natural frequency at which such a system will tend to oscillate: the stiffer the spring the higher the frequency of oscillation, and the higher the inertia of the flywheel the lower that frequency will be. This frequency is known as the resonant frequency.

We can drive this oscillator into larger and larger amplitudes of motion of the piston and the turbine by continuing to give the flywheel short pushes at the resonant frequency, much as one "pumps" a swing on a playground. You can easily see that the pressure developed between the top and the bottom of the capacitor will increase as the excursions of the piston increase; analogously, the voltage across a parallel LC circuit increases as the amplitude of the oscillating current increases.

Procedures

P1. Connect the sweep X-drive from the sweep generator to the X-axis input on the oscilloscope. On some units, these connections may be on the back. Connect the retrace blanking signal to the Z-axis or intensity input. Follow Figure 14.4.

P2. Connect a 50-ohm cable from the generator output to your LC circuit's input via clip leads. Connect the output to the other channel of the oscilloscope. This set-up will allow the 'scope to plot response versus frequency. Make sure the scope is in X-Y (or A versus B) mode.

P3. Draw schematics for a bandpass (allows a narrow frequency band to pass) and a notch filter (which is designed to block a particular frequency). Pick values for L, C, and the resonant frequency.

P4. Set up the frequency generator to sweep across a range of frequencies with your calculated resonant frequency approximately in the middle. Set up you components as a bandpass filter. Draw the response of your band-pass filter and locate the resonant frequency. Give an explanation of what is happening.

P5. Repeat P4 for a notch filter.

P6. Repeat either P4 or P5 for different component values.

References

P. Horowitz and W. Hill, *The Art of Electronics*, 2nd ed. (Cambridge, England: Cambridge University Press, 1989), pp. 41-2.

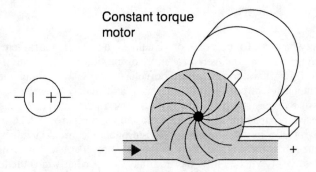

Figure 14.1   Water model for a steady voltage source (schematic symbol at left)

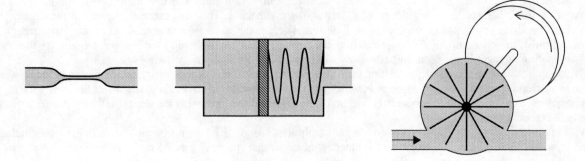

Figure 14.2   Water models and schematics symbols for a resistor (left), capacitor (center) and an inductor (right)

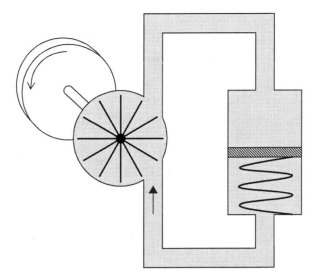

Figure 14.3 Water model for a parallel-LC circuit, with inductor on left and capacitor on right

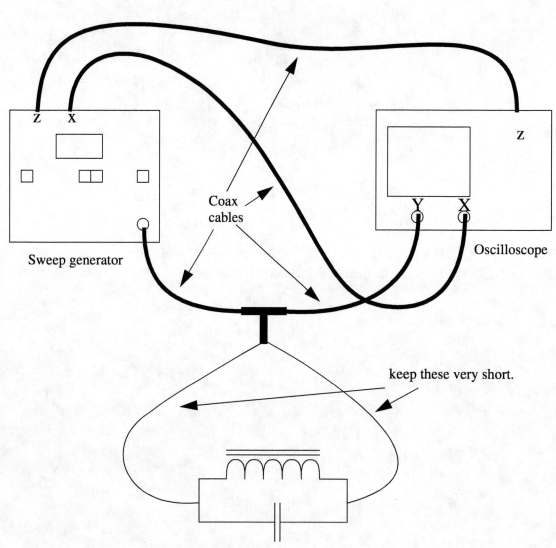

Figure 14.4 Experimental set-up

Resonant Filters

Questions

Q1. What formula determines the resonant frequency of the components? Where does this come from?

_____

Conclusions (What did you learn from this experiment?)

_____
_____
_____

Data and Observations

Signature: _____ Date: ___/___/___ Witness: _____ Date: ___/___/___

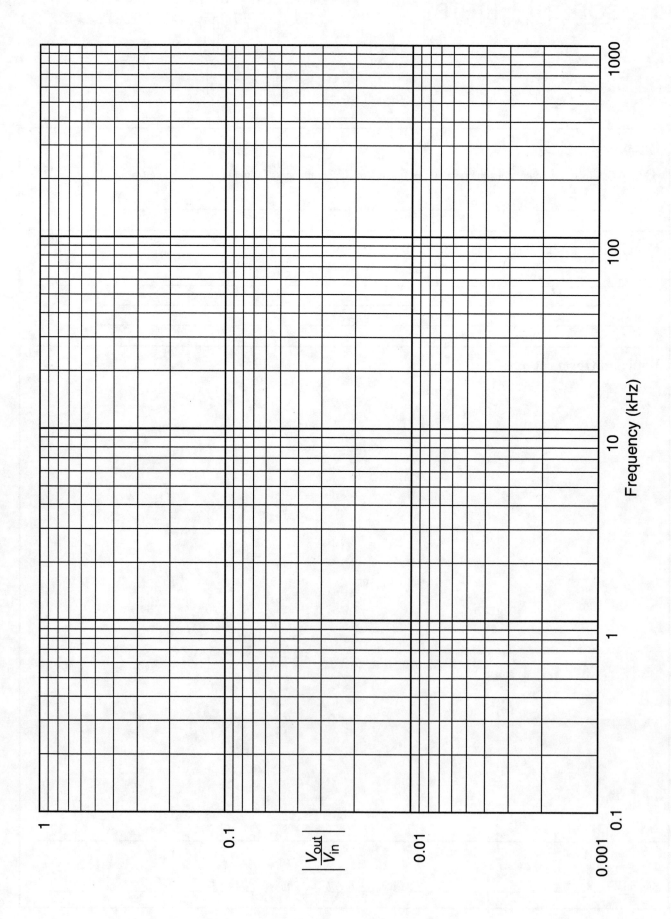

Frequency (kHz)

$\left| \dfrac{V_{\text{out}}}{V_{\text{in}}} \right|$

Prelab Questions: Resonant Filters

(Bring sheet with questions answered to your lab session)

Print your name (Last, First): _____

Q1. If the ordinary frequency, in Hz, of a sinusoidal signal is 100 Hz, what is the corresponding angular frequency of that signal (both magnitude and units)?

Q2. Give two examples of mechanical resonant devices.

Card Key

Instructional Objectives (At the end of this lab you should be able to:)

I1. Describe the construction of a resonant electrical circuit, and calculate its resonant frequency, given the values of its capacitance and inductance.

I2. Describe how the resonant card key is used.

Equipment

Passive card key (plastic encased printed circuit board with unique set of individually coded circuits); oscilloscope with 50 ohm input and X-Y mode; signal generator (to be swept from 2 to 20MHz); two meters of wire (to be wound into a coil); four 50-ohm cables; a 50 ohm T- connector; a BNC connector with test leads and a small piece of aluminum foil.

Description And Background

A fairly large business sector today is concerned with the security of buildings or individual rooms. A particular challenge is to unlock entrance ports when pre-approved individuals present a card key. In this lab we look at one type of practical card key that happens to employ a simple and important electronic component, the resonant circuit.

The card key we study is a passive device that looks rather like a conventional credit card. These cards, distributed by Schlage Electronics, contain no batteries, semiconductor circuits or magnetized strips. Instead they have imbedded in them four printed circuit coils (inductors) and four chip capacitors. (If you put the card against a bright light you may be able to see some of the internal circuitry.) The ends of each coil are connected to one of the chip capacitors, forming four independent parallel LC resonant circuits. The resonant frequency of each circuit on a given card is set at the factory to a different value. Employees of a particular company are issued cards that contain different sets of resonant frequencies. In other words, your frequencies are different from mine.

In a resonant circuit made by connecting an inductor, L, in parallel with a capacitor, C, energy will cycle back and forth between magnetic and electric fields. The magnetic energy storage occurs in the space occupied by and immediately surrounding the inductor, while the electric energy is stored in the dielectric between the plates of the capacitor. The energy exchange will occur at the so-called <u>resonant frequency</u>, which depends on the values of L and C.

The current and voltages in the circuit must satisfy both of the following equations:

$$v = L\frac{di}{dt} \text{ and } i = C\frac{dv}{dt}$$

which gives: $i(t) = I_0\sin\left(\frac{t}{\sqrt{LC}}\right)$ and $v(t) = V_0\cos\left(\frac{t}{\sqrt{LC}}\right)$ where $\frac{V_0}{I_0} = \sqrt{\frac{L}{C}}$

The formula for the resonant frequency is most simply given (and remembered) in terms of the angular frequency, $\omega_{res} = 2\pi f_{res}$. The formula is just $\omega_{res}^2 LC = 1$.

The idea of this experiment is to find the resonant frequencies of the card key that you test. To do this, you should wind a coil that is similar to inside the card key reader mounted outside a secured room. A ten-turn coil about ten centimeters across should work. You should drive the coil with a variable-frequency voltage obtained from a swept signal generator. Such a generator varies the frequency of its output repetitively between the extreme frequencies that you choose.

By monitoring the coil voltage with an oscilloscope, you can detect the resonant frequencies in the following way. As each resonant frequency is reached, energy is absorbed by the corresponding circuit on the card. This absorption of energy at the resonant frequency causes the coil voltage to decrease slightly. Measure and record those frequencies.

Note: You can perform this experiment with a signal generator whose frequency <u>you</u> "sweep" by hand; in this case, you must record the frequencies at which the generator output voltage drops. Alternatively, if the equipment is available, you can use an auto sweep generator; this device typically outputs the variable frequency and a "blanking" signal (often from an outlet on the back of the instrument). You should connect this blanking signal to the "Z" input of your oscilloscope (this connection may be on the back of the oscilloscope). The blanking signal will turn off the electron beam of the scope display when the frequency is being returned to its minimum (initial) value, producing a less confusing scope display.

Procedures

P1. Wind the long piece of wire into a circular coil about ten centimeters in diameter. Connect the coil to the test leads on the BNC connector and connect it to the T-connector. Connect a 50-ohm cable from the output of the variable frequency generator to the T-connector. Connect the oscilloscope to the remaining connection on the T with the remaining 50-ohm cable to monitor the voltage across the coil as the frequency is varied.

P2. Connect the sweep X-drive from the sweep generator to the X-axis input on the oscilloscope. On some units these connections may be on the *back*. There may also be a retrace blanking signal that can be connected to a Z-axis or intensity input

P3. Vary the frequency from 2MHz to 20MHz, observing the variation of voltage across the coil (the voltage should be relatively constant).

P4. Put the card key near the middle of the coil, orienting it parallel to the plane of the coil. Vary the frequency again and observe the variations of coil voltage. Measure and record the frequencies at which there are significant changes of the coil voltage (there should be four or five major variations). These are the resonant frequencies of this particular card.

P5. From the scope display determine the frequency difference, Δf_{3dB}, between the half-power points. Calculate the *quality factor:*

$$Q = \frac{f_{res}}{\Delta f_{3dB}}$$

The quality factor governs how many discrete resonant frequencies you could have in one card.

P6. While observing the scope display, bring a piece of aluminum foil into contact with one surface of the card key. (The electrically conducting foil reduces the space in which the magnetic field can exist around each of the printed inductors. This has the effect of reducing the inductance of each inductor and so raising the resonant frequency.)

References

P. Horowitz and W. Hill, *The Art of Electronics*, 2nd ed. (Cambridge, England: Cambridge University Press, 1989), pp. 41-2.

David Macaulay, *The Way Things Work* (Boston, MA: Houghton Mifflin, 1988), pp. 352, 322.

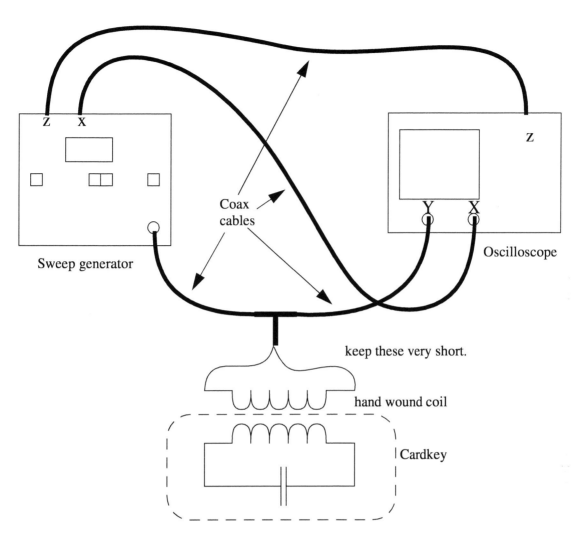

Figure 15.1 Lab setup. The cardkey houses multiple resonant circuits, each consisting of an inductor and a capacitor connected in parallel.

Card Key

Questions

Q1. How many unique card keys, of the type you just characterized, could be made?

_____

Conclusions (What did you learn from this experiment?)

_____
_____
_____

Data and Observations

Signature: _____ Date: ___/___/___ Witness: _____ Date: ___/___/___

Prelab Questions: Card Key

(Bring sheet with questions answered to your lab session)

Print your name (Last, First): _____

Q1. A card key contains several circuits that respond in a given frequency band. Each card key user has a unique set of response frequency bands. Suppose that each key has three such bands, each 0.5 MHz wide, and that a total of 20 MHz in all can be used. How many unique card key users could there be with such a system? (Assume that any frequency band is used only once on each card.)

Loudspeaker Crossover

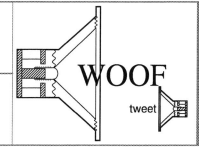

WOOF
tweet

Instructional Objectives (At the end of this lab you should be able to:)

I1. Describe the circuitry and function of a loudspeaker crossover network.

Equipment

Oscilloscope and probe; signal generator with frequency sweep; stereo receiver or amplifier with equalization adjustment and tape or CD input; high fidelity speakers, one of which has been modified to make the connections to the individual loudspeakers accessible; dynamic microphone; tape deck with microphone jack; two BNC-to-RCA adapter; four 50-ohm cables.

Description and Background

Loudspeakers are often designed to provide exceptional performance in a given frequency range. A typical high-fidelity speaker system will use three separate loudspeakers in a single enclosure. Therefore, circuitry is needed between the output of a music system and the loudspeakers that "sends" signals of the proper frequencies to each loudspeaker. Such a circuit is known as a <u>crossover network</u>.

Procedures

P1. Connect the sweep X-drive from the sweep generator to the X-axis input on the oscilloscope. (On some units, these connections may be on the back of the instrument.) Connect the retrace blanking signal (described in the cardkey lab) to the Z-axis (intensity modulation) input. Follow Figure 16.1.

P2. Connect a 50-ohm cable from the generator output to the tape or CD input on the stereo receiver using the adapter. Connect the oscilloscope probe to one of the individual speaker connections provided on the modified speaker cabinet.

P3. Set the function generator so that it sweeps over the audio range. Start with a low voltage and raise it as needed. Don't forget that you have a volume control on the amplifier if you can't hear anything! Try different sweep ranges to get an idea of the type of filtering being used.

P4. Try adjusting the treble and bass knobs on the amplifier. Sketch what happens.

P5. Repeat P3 and P4 using the other speaker connections.

P6. Replace the oscilloscope probe in the setup with the tape deck and the microphone. Set the tape decks controls so that you get a signal from the microphone on its line output. Usually you must either put a tape into the deck, or depress the write protect switch to fool the deck, and set it to record. adjust the record levels to mid-range. Hold the microphone in front of each speaker element at close range, while sweeping through appropriate frequencies. Draw the response envelope for each speaker.

P7. Estimate the 3dB cutoffs for the crossover circuits. What kinds of filters are they?

References

David Macaulay, *The Way Things Work* (Boston, MA: Houghton Mifflin, 1988), pp. 240-241.
See "Transfer Functions for Basic R-L and R-C Filter Circuits" on page 168.

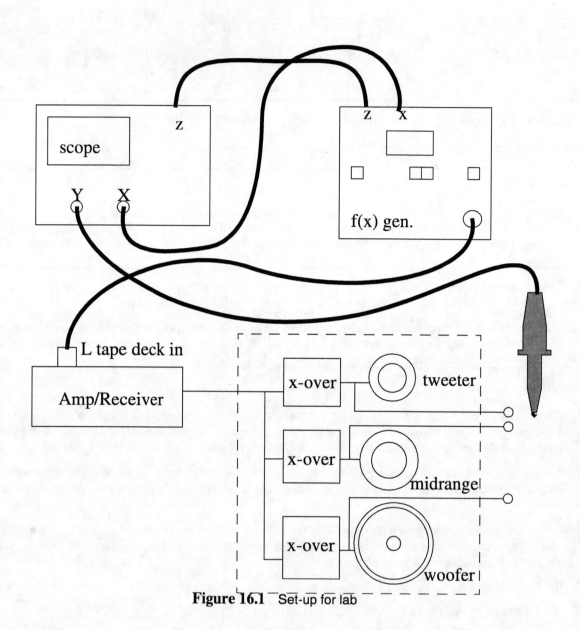

Figure 16.1 Set-up for lab

Loudspeaker Crossover

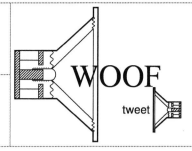

Questions

Q1. Why is the measured audio output of each speaker different from the electrical drive signal supplied to it?

_____

_____

_____

Conclusions (What did you learn from this experiment?)

_____

_____

_____

Data and Observations

Signature: _____ Date: ___/___/___ Witness: _____ Date: ___/___/___

Prelab Questions: Loudspeaker Crossover

(Bring sheet with questions answered to your lab session)

Print your name (Last, First): _____

Q1. A typical tweeter (high-frequency loudspeaker) has a two-inch diameter. How large is this in terms of wavelengths of sound at a nominal operating frequency of 8 kHz? (Assume that the velocity of sound in air is 1090 feet per second.)

_____
_____
_____

Q2. Do you think the frequency range sent to each speaker should have sharp boundaries? Why?

_____
_____
_____
_____

AM Radio Transmitter

Instructional Objectives (At the end of this lab you should be able to:)

I1. Describe how AM radio works.

Equipment

Science Fair 130-in-one Electronic Project Lab; oscilloscope; audio signal generator; pocket radio; headphones.

Description And Background

The radio and television signals that you receive in your car or dwelling are produced in essentially the same ways (see Figure 17.1). The information to be transmitted (speech or music from a microphone, or pictures from a video camera) is used to <u>modulate</u> (change) a high-frequency sinusoidal voltage called the <u>carrier</u>. The resultant voltage is <u>amplified</u> and used to drive an <u>antenna</u> that radiates electromagnetic waves into the surrounding air. When a portion of this radiated electromagnetic wave reaches your receiver's antenna, it produces a voltage at the antenna's terminals.

To recover the information carried on this wave, the receiver <u>selects</u> the desired station by <u>filtering</u> its signal from all those reaching the antenna. The filter is set at the station's known carrier frequency (such as "1610 AM", "107.7 FM", "Channel 2"). The selected signal is now <u>demodulated</u>, a step in which the frequencies representing the information are separated out. These then go to an <u>amplifier</u> and to a loudspeaker, or to a picture display device.

In this experiment, the information coding scheme that you will use is called <u>amplitude modulation</u>, or AM. The information to be transmitted causes the <u>amplitude</u> of the carrier-frequency voltage to vary with time. In a later course you may analyze this modulation process (as well as its companion process, frequency modulation, or FM) in detail.

The transmitter you will build, using the *Science Fair 130-in-one Electronic Project Lab*, employs a loudspeaker as a microphone (you might ask your instructor about how this works). The voltage produced is then amplified and used to control the amplitude of the output of an LC resonant circuit: this is the amplitude-modulated transmitter output.

Procedures

P1. Build the "AM Radio Station" using the Science Fair 130-in-one Electronic Project Lab. Connect the oscilloscope probe to the far end of the antenna (the very long green wire). Use the positive battery terminal for ground. Set the pocket radio to receive at 1 MHz. Optimize the tuning of your transmitter using feedback squeal from the radio, moving it away to fine tune it. Draw the schematic.

P2. Connect an audio signal generator to your transmitter instead of the speaker/transformer. Draw waveforms of no modulation, full modulation, and over modulation at audio frequencies. Identify the carrier and the envelope on your drawings.

P3. Try to figure out the function of each transistor.

References

P. Horowitz and W. Hill, *The Art of Electronics*, 2nd ed. (Cambridge, England: Cambridge University Press, 1989), pp. 885-6, 894-6.

David Macaulay, *The Way Things Work* (Boston, MA: Houghton Mifflin, 1988), pp. 254-5.

Dick White and Roger Doering, *Electrical Engineering Uncovered* (Upper Saddle River, N.J.: Prentice Hall, 1997), "Radio in a Nutshell" on page 240.

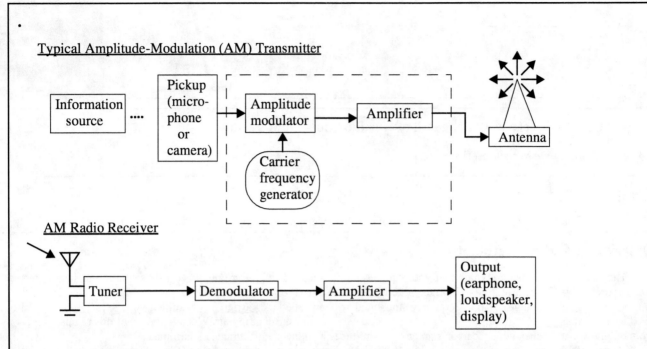

Figure 17.1   Block diagrams of an amplitude-modulation (AM) radio transmitter (top) and an AM radio receiver (bottom). The functions in the dotted rectangle are arranged differently in this lab.

AM Radio Transmitter

Questions

Q1. What is overmodulation?

_____

Q2. Why is there still a carrier wave when there is no modulating signal?

_____

Conclusions (What did you learn from this experiment?)

_____
_____
_____

Data and Observations

Signature: _____ Date: ___/___/___ Witness: _____ Date: ___/___/___

Prelab Questions: AM Radio Transmitter

(Bring sheet with questions answered to your lab session)

Print your name (Last, First): _____

Q1. The figure below shows a fully modulated (100%) AM signal. Sketch a 0% modulated version and a 50% modulated version. (The text above the figure is the Mathcad input required to produce the figure.)

Modulation percentage $\quad$ P $:= 100 \cdot \%$
Modulation frequency $\quad$ M $:= 1 \cdot kHz$
Carrier frequency $\quad$ C $:= 10 \cdot kHz$
Modulating signal $\quad$ Mod(t) $:= P \cdot \sin(2 \cdot \pi \cdot M \cdot t)$
AM signal $\quad$ AM(t) $:= (1 + Mod(t)) \cdot \cos(2 \cdot \pi \cdot C \cdot t)$
$\quad$ t $:= 0 \cdot sec, 0.005 \cdot msec .. 4 \cdot msec$

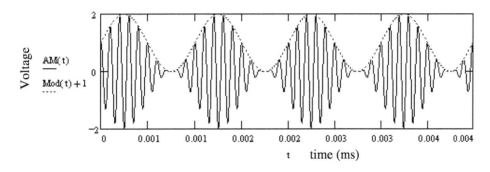

AM Radio Receiver

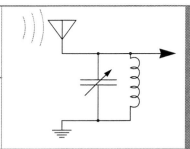

Instructional Objectives (At the end of this lab you should be able to:)

I1. Explain how a receiver converts an amplitude modulated radio carrier into an audible signal.

Equipment:

Analog oscilloscope and two probes; signal generator, with at least one MHz range, with internal or external amplitude modulation capability; audio signal generator or audio source; Science Fair 130-in-one Electronic Project Lab.

Description and Background

First review the "Description and Background" for the lab "AM Radio Transmitter" on page 125.

The simplest AM radio receiver is what was once known as a "crystal set". You will first build and test such a receiver, using parts in your Science Fair 130-in-one Electronic Project Lab. Then you will add a transistor amplifier to the crystal set to increase the audibility of the radio signal.

The crystal set has no batteries, transistors, or vacuum tubes. Just like the crystal sets that individuals used in the very early days of radio, this set is powered entirely by the energy radiated from the transmitter's antenna. (Instructor's Note: Depending upon the shielding in your lab room, you may need to set up a low-power radio transmitter in the room in order to have an audible output from the crystal set. This can be done by assembling the Science Fair 130-in-one Electronic Project Lab's "AM Radio Station" and using either an audio tone generator or a cassette player output as the modulation source. Alternatively a radio frequency signal generator with A.M. input can be used as the "transmitter.")

The crystal set circuit is shown in Figure 18.1. The electric field at your antenna causes charges to move up and down between the top of the antenna and ground, generating a magnetic field in the coil. The coil and the variable capacitor form a filter; the resonant frequency of the filter should be set to the carrier frequency of the AM radio station you want to pick up. When this condition exists, the voltage developed across the variable capacitor is maximized. This voltage is rectified by the diode, producing a sequence of unipolar (zero to positive polarity) pulses, much like the output of the unfiltered half-wave rectifier discussed in *Electrical Engineering Uncovered,* on page 158. As with that rectifier, the fixed-value capacitor that follows charges up to the peak values of the voltage pulses. If the time constant of the RC circuit to the right of the diode is properly chosen, the voltage across the terminals of the earphone will be a steady DC term plus the modulation — the speech or music impressed on the carrier in the modulation stage of the AM transmitter.

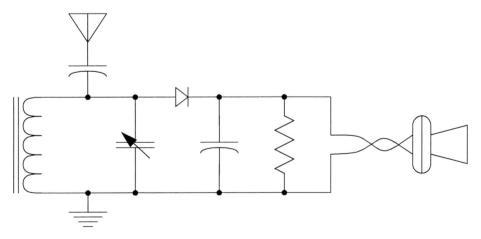

Figure 18.1 Crystal set radio (Simple-Diode Radio) from the *Science Fair 130-in-one Electronic Project Lab*

Procedures:

P1. Read the text associated with the crystal radio of kit #110. Build it.

P2. Do not be alarmed if your assembled kit does pick up any stations, or even static. The building may act as a shield so you would have to take your kit outside to pick up anything (if your instructor permits it). Note that the receiver/amplifier and the TV in our lab are both connected to an antenna on the roof of the building.

P3. Since you may not be able to pick up any stations in the lab, you can use the frequency generator as your transmitting station. Set the generator to produce a modulated radio signal around 1MHz. This is about the middle of the AM band, and your receiver should be easily tuned to this station. Set the amplitude at 30mV. Set the internal [modulating] signal to a sine wave, or apply an external sine wave or other audio signal to modulate the carrier.

P4. What is the function of the "tank circuit"? Identify the tank circuit in the schematic. Disconnect the variable capacitor without changing its setting. Measure its capacitance. Use this information to estimate the value of the inductor in the kit. See the lab "Card Key" on page 111 for the resonant frequency calculation.

P5. After experimenting with the crystal set to see that it actually works, you are ready to add the two-stage transistor amplifier and switch from the earphone to a loudspeaker (see circuit in the Science Fair 130-in-one Electronic Project Lab for the "Two-Transistor Radio"). (Note that you should also change the values of the fixed capacitor and the resistor to those used in the circuit for the two-transistor radio.)

P6. Build the Science Fair 130-in-one Electronic Project Lab on page 134 (kit #111): "Two Transistor Radio". Draw its schematic. Connect your antenna to the output of the RF signal generator. You may have better luck using a short wire. Tune your radio to hear a tone now from your radio's speaker.

P7. Using the positive battery terminal as ground, look at the signal that comes in from the antenna. This should look familiar from the AM Transmitter lab. Try changing the modulation frequency. What happens? Now, try changing the percentage of modulation, or the amplitude of the modulating signal. Try 100%. Sketch what happens. These should also look familiar.

P8. Sketch the signal on terminal 125, after the diode. How does this compare to the signal on the antenna? Why?

P9. Sketch the signal after transistors Q1 and Q2. What functions do these transistors perform?

References

Read "Crystal Set Radio (Simple-Diode Radio)" and "Two Transistor Radio" in the *Science Fair 130-in-one Electronic Project Lab* (Fort Worth, TX: RadioShack [Tandy Corp], 1989)

P. Horowitz and W. Hill, *The Art of Electronics*, 2nd ed. (Cambridge, England: Cambridge University Press, 1989), pp. 894-6.

AM Radio Receiver

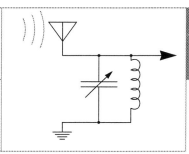

Questions

Q1. What is the 3 dB frequency of the filter after the diode detector? Why?

_____

Conclusions (What did you learn from this experiment?)

_____
_____
_____

Data and Observations

Signature: _____ Date: ___/___/___ Witness: _____ Date: ___/___/___

Prelab Questions: AM Radio Receiver

(Bring sheet with questions answered to your lab session)

Print your name (Last, First): _____

Q1. If you hung up a long wire on an insulating wooden pole, and connected the lower end to the input of your oscilloscope, it seems that you should pick up radio signals from any nearby or powerful transmitter. In what ways do you think this radio "receiver" would be somewhat deficient?

_____

_____

_____

_____

_____

_____

_____

Operational Amplifiers

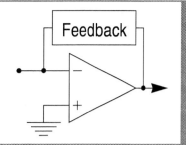

Feedback

Instructional Objectives (At the end of this lab you should be able to:)

I1. Design inverting and non-inverting amplifiers using operational amplifiers.

Equipment

Science Fair 130-in-one Electronic Project Lab; condenser microphone element (RadioShack cat.no. 270-090); clipleads; oscilloscope.

Description and Background

Operational amplifiers are used as basic building blocks when designing analog circuits. They provide a nearly ideal gain block with high input resistance, low output resistance and extremely high open-loop gain. Please read about op-amps in *Electrical Engineering Uncovered*, "Operational Amplifiers (Op-Amps)" on page 180

Procedures

P1. Build the circuit shown in Figure 19.1. Use clip-leads to attach the microphone to terminals 13 and 14. Please observe that the ground lead of the microphone element has a metal trace connecting to the case. The terminal marked with a ground sign in the op-amp section of the kit board is NOT ground, but rather the negative power supply connection for the op-amp.

P2. Check out the circuit by humming, speaking, singing or whistling into the microphone, while observing the output on the oscilloscope.

P3. Connect the output of the amplifier to the transformer and the output of the transformer to the speaker.

P4. Modify the circuit to use the potentiometer (control) to allow you to change the gain.

References

P. Horowitz and W. Hill, *The Art of Electronics*, 2nd ed. (Cambridge, England: Cambridge University Press, 1989), pp. 175-85.

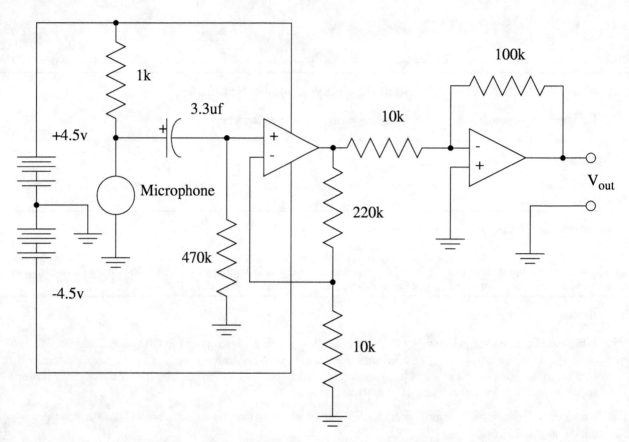

Figure 19.1 Microphone amplifier

Op Amps

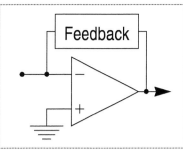

Questions

Q1. Should an op-amp be used if the desired gain is less than one?

_____

Conclusions (What did you learn from this experiment?)

_____

_____

_____

Data and Observations

Signature: _____ Date: ___/___/___ Witness: _____ Date: ___/___/___

Prelab Questions: Op-Amp

(Bring sheet with questions answered to your lab session)

Print your name (Last, First): _____

Q1. Please answer the following true-false questions, explaining your answers:

a. Op-amps can be located easily in a box of integrated circuits because they have a triangular shape.

b. An op-amp is a semiconductor device that is unique because it uses neither electrons nor holes.

Multiband Transmitter

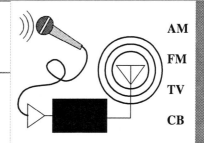

AM
FM
TV
CB

Lab 20

Instructional Objectives (At the end of this lab you should be able to:)

I1. Describe how FM radio works.

I2. Calculate spectrum utilization (the bandwidth that a channel requires).

Equipment

RF signal generator with external modulation; microphone; tape deck with microphone input, or mixer with microphone input or the completed op amp lab circuit; oscilloscope; AM/FM receiver; citizens band radio ("walkie talkie"); television; 50-ohm cables and BNC-to-RCA adapter; optional RF spectrum analyzer and audio signal generator.

Description and Background

We explored amplitude-modulate (AM) radio in the AM transmitter lab. AM was the first type of modulation to carry voice and music. AM radio has excellent long distance propagation characteristics, especially at night when its electromagnetic waves with frequencies in the AM band will bounce off the ionosphere. However, AM suffers from mixing, when two stations are using the same carrier, as their signals will mix in proportion to the received signal strengths, and from interference from lightning and other electronic noise.

Citizens band radio, used for short range personal communication, also utilizes AM.

Frequency modulation (FM) combats the problems of interference, and is much more immune to noise. Its weakness is that when a mobile receiver is in an area with many reflecting surfaces, it will suffer noisy transitions from one image (or reflection) to the next as the strength of the reflections changes due to the motion of the receiver.

The higher quality that we experience on the broadcast FM band is also due to the large bandwidth that has been allocated for each channel.

Stereo is achieved by adding the left and right signals for the base modulating signal, then the difference between the two channels is shifted in frequency up to 38kHz. A pilot tone of 19 kHz is added to allow the receiving station to double it to demodulate the second channel.

Television also uses FM to transmit its audio component in a portion of the allocated band.

Procedures

P1. Connect the microphone to the tape deck, set it to record (it can be on "pause"). On most tape recorders you can defeat the switch, which is looking for the record enable tab on the cassette, by simply pressing the switch arm while pressing the record button. Alternatively, use a mixer or the op-amp lab setup to get a 1-volt audio signal. Verify your signal level with the oscilloscope.

P2. Connect the audio signal to the external modulation input of the RF signal generator. Set the generator for external modulation. Connect the output to the spectrum analyzer and to a six-foot piece of wire (to act as an antenna), or wire it directly to the antenna terminal on the receiver. Set the amplitude at 30mV (**don't overload the receiver inputs**).

P3. For each row of the following table find an unused frequency in your locality and do the remaining procedures (avoiding CB channel 9, the emergency channel):

Modulation Type	Radio	Carrier Frequency			
		min	step	max	unit
AM	AM	530	10	1620	kHz
FM	FM	87.9	0.2	107.9	MHz
AM	CB	26.965	0.01	27.405	MHz
FM	TV(2-6)	59.75	6	87.75	MHz

P4. Tune the appropriate receiver to an unused channel frequency. Set the RF signal generator to a matching frequency. You should be able to talk into the mike now and hear yourself on the receiver.

P5. Set the spectrum analyzer to show a 40 kHz band around your carrier frequency. Sketch what you see when talking.

P6. Substitute the audio signal generator for the external modulating signal. Sketch the spectrum with the audio generator set at 0.1Hz, 1kHz, 10kHz and 20kHz.

P7. Reduce the amplitude of the signal generator output. What happens? There should be a point at which static overcomes voice signal. Record this amplitude.

P8. Vary the modulation percentage (AM) or the deviation in kHz (FM). What happens?

P9. Switch the modulation type (AM ↔ FM). Does the receiver still work? Speculate on why.

References

P. Horowitz and W. Hill, *The Art of Electronics*, 2nd ed. (Cambridge, England: Cambridge University Press, 1989), pp. 898-9.

Robert Schetgen, ed., *Handbook for Radio Amateurs*, 70th ed. (Newington, CT: American Radio Relay League, 1993).

Gordon McComb, *Troubleshooting & Repairing VCRs*, 2nd ed. (Blue Ridge Summit, PA: TAB Books, 1991), pp 375.

Dick White and Roger Doering, *Electrical Engineering Uncovered* (Upper Saddle River, N.J.: Prentice Hall, 1997), "Waveforms and Spectra" on page 235.

David Macaulay, *The Way Things Work* (Boston, MA: Houghton Mifflin, 1988), p. 254.

Multiband Transmitter

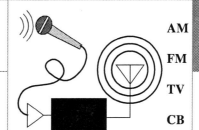

AM
FM
TV
CB

Questions

Q1. Why is the bandwidth of a radio signal important?

_____

Conclusions (What did you learn from this experiment?)

_____

_____

_____

Data and Observations

Signature: _____ Date: ___/___/___ Witness: _____ Date: ___/___/___

Prelab Questions: Multiband Transmitter

(Bring sheet with questions answered to your lab session)

Print your name (Last, First): _____

Q1. The following figure illustrates a frequency modulated carrier and the single-frequency sinusoidal modulating signal. The carrier frequency is 30 kHz, the modulating signal is one kHz and the frequency deviation is 15 kHz. The time is in milliseconds. The text above the figure is the input to Mathcad, the program which generated the figure.

Modulation Bandwidth $M := 15 \cdot kHz$

Modulating frequency $f := 1 \cdot kHz$

Modulating signal $s(t) := \cos(2 \cdot \pi \cdot f \cdot t)$

Carrier frequency $c := 30 \cdot kHz$

FM signal $FM(t) := \cos\left(2 \cdot \pi \cdot c \cdot t + 2 \cdot \pi \cdot M \cdot \int_{0 \cdot sec}^{t} \cos(2 \cdot \pi \cdot f \cdot x) \, dx\right)$

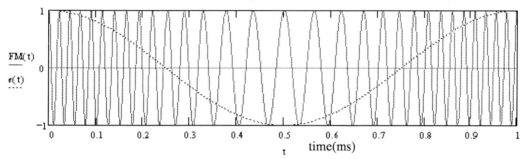

What is the maximum frequency in the waveform?_____

The minimum?_____

What is the total bandwidth of this signal?_____

Radio-Controlled Car

Instructional Objectives (At the end of this lab you should be able to:)

I1. Figure out the coding scheme of the radio-controlled car signals.

I2. Enjoy the fruits of your labor — take the car outside and run it around!

Equipment

Radio-controlled car with charged battery; oscilloscope (preferably digital) and probe.

Description and Background

Many different coding schemes could be used to control a radio-controlled car, even one that has a very limited repertoire — go forward, go backward, turn left, turn right.

In previous labs you have studied amplitude and frequency modulation (AM, FM). You could also signal with pulses whose widths are varied to convey information (pulse width modulation, or PWM). Alternatively, you could vary the time interval between successive pulses (pulse position modulation, PPM). The aims of this lab are to discover which type of coding is used, and figure out how it is used... and then go play with the car.

Procedures

P1. Connect an oscilloscope probe to the antenna of the controller, connect a clip-lead to one of the battery terminals in the battery compartment to serve as the oscilloscope probe's ground, and turn the controller "on". (Don't turn on the car yet.)

P2. Obtain a triggered signal on the screen that changes obviously when the controller switches are manipulated. You will probably need to adjust either the trigger holdoff, or the sweep vernier to achieve stable triggering.

P3. Record what happens to the signal as the controls are moved to command forward, backward, left and right. Sketch the waveforms; be sure to include time information in your sketches.

P4. Find the carrier frequency that is used.

P5. Determine which part of the signal is used for going forward and backward, and which for left and right.

P6. Determine what type of coding (AM, FM, PWM, or PPM) is used.

P7. The fun part: go play with the car (but don't disturb people in nearby classrooms).

References

P. Horowitz and W. Hill, *The Art of Electronics*, 2nd ed. (Cambridge, England: Cambridge University Press, 1989), pp. 900-1.

Radio-Controlled Car

Questions

Q1. Answer the questions implied in Procedures P4, P5 and P6.

Conclusions (What did you learn from this experiment?)

_____

_____

_____

Data and Observations

Signature: _____ Date: ___/___/___ Witness: _____ Date: ___/___/___

Prelab Questions: Radio-Controlled Car

(Bring sheet with questions answered to your lab session)

Print your name (Last, First): _____

Q1. Assume that you wanted to "tell" the car to move <u>forward</u> at <u>half speed</u>. Choose whichever of the four modulation schemes that you wish, and sketch waveforms that might instruct the car to move as stated. Write out any assumptions you make about the coding scheme that you choose.

CD Player

Compact Disc

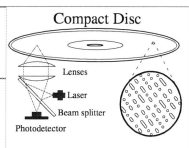

Lenses

Laser

Beam splitter

Photodetector

Lab

22

Instructional Objectives (At the end of this lab you should be able to:)

I1. Describe the coding used on compact discs.

Equipment

Specially modified CD player; amplifier and two speakers; oscilloscope; two 50-ohm cables or probes; "The Ultimate Test CD"; sound pressure level meter.

Description and Background

Compact discs have swept the music and computer markets. They provide a high capacity (700 Mbyte) and therefore long playing medium for the storage of music and data.

Before music can be stored on this digital medium it must be digitized. Each channel of the source is sampled, and the analog value converted to a digital code using 16 bits. Assuming that the digitization is nearly perfect (that is, that each value sampled is represented by the nearest digital code), then each code contains an error no greater than one-half the value of the least significant bit. This means the swing in the error from plus one-half to minus one-half is one bit. Thus the signal-to-sampling-noise ratio for a full scale signal is:

$$20\log\frac{2^{16}}{1} = 96\text{dB}$$

While there is one signal recorded at full scale on the test CD, most music is not recorded at such levels in order to leave headroom for sudden louder signals. Also, many classical recordings feature passages that are extremely quiet. Under these situations, the signal-to-noise ratio declines.

Sound levels on the disc are referred to the full-scale value, which becomes zero dB. All levels are thus below zero.

See "Bit Rate" on page 28 and "CD" on page 249 in the text.

Procedures

P1. Check (and correct if necessary) the speaker and CD player wiring to ensure that left signals are coming from the left speaker, and right signals from the right speaker.

P2. Check (and correct if necessary) the phasing of the speakers (or headphones) using the pink noise tracks. The in-phase pink noise should appear to come from between the speakers (or inside your head with headphones). Out-of-phase noise will appear to come from behind you or from no particular direction. If the speakers are out of phase, these effects will be reversed. To correct the speaker phasing, reverse the wires on one speaker only.

P3. Check your own hearing range using the progressively higher frequency tracks. Set the volume at a comfortable level (no more than 80 dB) on the 1kHz track. If you're using head phones, measure the level right in the ear-piece. **Do not increase the volume above this level on higher frequency tracks as you may damage your hearing**. Depending on your player you may hear artifacts of poor filtering on the 20kHz track.

P4. If you can, remove the speaker grill so that you can watch the cone of the woofer. Play the 20 Hz track. You may need to increase volume levels somewhat to see the motion of the cone. Too much power can pop the speaker's circuit breaker, or, if it is unprotected, damage the voice coil of the woofer. If the speaker has a tuned port (a hole into the enclosure) place your hand in front of it to feel wind.

P5. The player for this lab has been specially modified to bring out the digital bit stream, right before it enters the D/A converters. There are two signals which have been brought out. Connect them to the oscilloscope. Observe the waveforms while playing silence, left only, right only and music. Sketch the waveforms. Determine what information is on each wire.

P6. Determine the bit rate and the sample rate per channel. These have already been oversampled (see "Sampling, Filters, Playback and All That........." on page 250 of the text). Determine the sampling rate used on the disc given the oversampling rate of your player.

P7. Verify your sampling rate by estimating the number of samples recorded on the entire CD (70 minutes and 700 megabytes).

P8. Sketch five cycles of a 20 kHz sine wave. Mark the sampling points a CD would use on your sketch.

References

P. Horowitz and W. Hill, *The Art of Electronics*, 2nd ed. (Cambridge, England: Cambridge University Press, 1989), p. 676.

David Macaulay, *The Way Things Work* (Boston, MA: Houghton Mifflin, 1988), pp. 243; 248-9; 340.

Dick White and Roger Doering, *Electrical Engineering Uncovered* (Upper Saddle River, N.J.: Prentice Hall, 1997), "Bit Rate" on page 28 and "Decibel Measure for Sound Pressure" on page 25.

Instruction brochure accompanying "The Ultimate Test CD".

CD Player

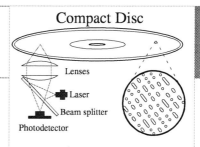

Compact Disc

Lenses
Laser
Beam splitter
Photodetector

Questions

Q1. What sampling rate is adequate for the audio spectrum (20-20,000 Hz)?

_____

Conclusions (What did you learn from this experiment?)

_____

_____

_____

Data and Observations

Signature: _____ Date: ___/___/___ Witness: _____ Date: ___/___/___

Prelab Questions: CD Player

(Bring sheet with questions answered to your lab session)

Print your name (Last, First): _____

Q1. What is the minimum number of samples per second need to reproduce a 1,000 Hz musical note?

_____

Half Adder

Instructional Objectives (At the end of this lab you should be able to:)

I1. Connect logic gates to perform a desired function.

Equipment

Science Fair 130-in-one Electronic Project Lab.

Description and Background

The function of the half adder is to produce the sum and carry bits from two input bits. The sum is the XOR of the two inputs, and the carry is the AND of the two inputs.

The experiment kit has a project to build an XOR using the four NAND gates supplied.

Procedures

P1. Wire up the XOR as shown in the manual for the *Science Fair 130-in-one Electronic Project Lab*, using four NAND gates. Verify that the output, shown on the LED, indicates the sum of the inputs.

P2. Add another LED (making sure to use the adjacent resistor to limit current) to the circuit to indicate the carry output. This is the tricky part of this lab.

References

Dick White and Roger Doering, *Electrical Engineering Uncovered* (Upper Saddle River, N.J.: Prentice Hall, 1997), "Digital Logic Devices" on page 201.

Half Adder

Half $\underset{\text{\large)}}{\overset{\text{\large 5}}{\underline{+\ 4}}}$ Adder

Conclusions (What did you learn from this experiment?)

_____
_____
_____

Data and Observations

Signature: _____ Date: ___/___/___ Witness: _____ Date: ___/___/___

Prelab Questions: Half Adder

(Bring sheet with questions answered to your lab session)

Print your name (Last, First): _____

Do one of these:

Q1. Draw a schematic of an XOR function implemented with only NAND gates.

Q2. Draw a schematic of how you can light an LED with the output of a logic gate when its value is logical zero. You can use a resistor and connections to the power supply as necessary.

Simulink

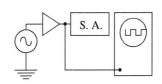

Instructional Objectives: (At the end of this lab you should be able to:)

I1. Construct simple systems using Simulink and understand the functions of system components (such as a gain block).

I2. Deswcribe what is meant by "transfer function".

I3. Describe whatis meant by the frequency content of a signal.

Equipment

Personal computer with Simulink installed.

Description And Background

Simulink is an addition to the MATLAB software package. Simulink allows you to construct systems (in block diagram form) and run simulations on them. You will be using Simulink to understand the function of a gain block, to examine the frequency content of various signals (like a square wave and sine wave), and to construct a lowpass filter and examine its frequency response.

This notation involves a complex-valued quantity denoted by the letter **s**. All that's important now is to see how **s** relates to something you're already somewhat familiar with — the frequency of a time-varying voltage or current.

Recall that we customarily represent an ac voltage as a periodic function of time such as

$$V(t) = V_0 \cos(\omega \tau) \tag{1}$$

where V_0 is the amplitude of the voltage, t is time, and ω is the so-called angular frequency, whose units are radians per second. The angular frequency is related to the "ordinary" frequency, f, measured in Hertz, by

$$\omega = 2\pi f \tag{2}$$

For example, if the frequency, f, of the ordinary powerline voltage is 60 Hz, then the associated angular frequency, ω, is 377 radians/s ($2\pi \times 60$).

For now, we'll consider the parameter **s** to be a purely imaginary quantity whose magnitude equals the angular frequency ω divided by a normalizing angular frequency ω_N. In other words, for our purposes,

$$\mathbf{s} = \mathbf{j}(\omega/\omega_N) \tag{3}$$

where $j = (-1)^{1/2}$. (We've written **s** in bold-faced type to emphasize that it is a complex-valued, rather than real, quantity.)

v_{in} v_{out}

Figure 24.1 Two-port circuit

The transfer function (ratio of output voltage to input voltage) for a two-port circuit is expressed in SIMULINK as a function of **s**, not as a function of ω or f. But by using Eq. (3), you can convert from an equation involving **s** to one simply involving angular frequency. For example, suppose that the transfer function for some circuit (Figure 24.1) is expressed in SIMULINK as

$$\mathbf{V}_{out}/\mathbf{V}_{in} = 1/(\mathbf{s} + 1) \tag{4}$$

The complex quantities $\mathbf{V}_{in}$ and $\mathbf{V}_{out}$ relate to the real-world input and output voltages v_{in} and v_{out}, which are real functions of time. (You'll learn in a later course why using these complex voltages to represent the real-world voltages is helpful when analyzing circuits.) Following the rules for finding the magnitude of a complex quantity, you arrive at an expression for the magnitude of the transfer function in terms of frequency as follows:

$$\left|\frac{\mathbf{V}_{out}}{\mathbf{V}_{in}}\right| = \left|\frac{1}{\mathbf{s}+1}\right| = \frac{1}{\sqrt{\mathbf{s}\mathbf{s}^* + 1}} = \frac{1}{\sqrt{(\omega/\omega_N)^2 + 1}} \tag{5}$$

where $\mathbf{s}^*$, the complex conjugate of $\mathbf{s}$, is just $\mathbf{s}^* = -j(\omega/\omega_N)$.

To see how this works for simple R-L and R-C circuits, we list below the transfer functions for four fundamental circuits that you might explore now in terms of angular frequency. We also give the transfer function in terms of $\mathbf{s}$, as well as a sketch of the magnitude of the transfer function versus the normalized angular frequency, ω/ω_N.

In order to simulate one of these circuits in SIMULINK, enter the corresponding $\mathbf{s}$-function and interpret the results in terms of frequencies by using the proper expression for the normalizing frequency ω_{RL} or ω_{RC} as given below.

Transfer Functions for Basic R-L and R-C Filter Circuits

SERIES L, PARALLEL R CIRCUIT

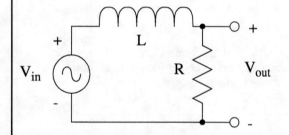
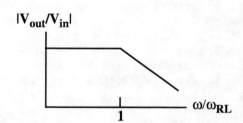

$$\left|\frac{V_{out}}{V_{in}}\right| = \frac{1}{\sqrt{(\omega L/R)^2 + 1}} \text{, or } \left|\frac{V_{out}}{V_{in}}\right| = \frac{1}{\sqrt{(\omega/\omega_{RL})^2 + 1}} \text{, where } \omega_{RL} = R/L. \tag{6}$$

In $\mathbf{s}$ notation, $V_{out}/V_{in} = 1/(\mathbf{s}_{RL} + 1)$, where $\mathbf{s}_{RL} = j(\omega/\omega_{RL})$.

SERIES R, PARALLEL L CIRCUIT

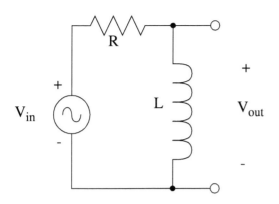

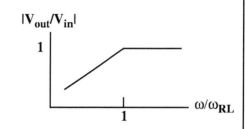

$$\left|\frac{V_{out}}{V_{in}}\right| = \frac{(\omega L)/R}{\sqrt{((\omega L)/R)^2)+1}} \quad , \text{or} \quad \left|\frac{V_{out}}{V_{in}}\right| = \frac{(\omega/\omega_{RL})}{\sqrt{((\omega L)/R)^2+1}} \tag{7}$$

where $\omega_{RL} = R/L$.

In **s** notation, $V_{out}/V_{in} = s_{RL}/(s_{RL}+1)$, where $s_{RL} = j(\omega/\omega_{RL})$.

SERIES C, PARALLEL R CIRCUIT

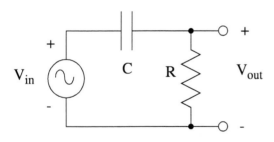

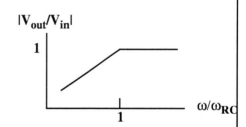

$$\left|\frac{V_{out}}{V_{in}}\right| = \frac{\omega \dot R C}{\sqrt{(\omega R C)^2+1}} \quad \text{or} \quad \left|\frac{V_{out}}{V_{in}}\right| = \frac{(\omega/\omega_{RC})}{\sqrt{(\omega/\omega_{RC})^2+1}} \tag{8}$$

where $\omega_{RC} = 1/RC$

In **s** notation, $V_{out}/V_{in} = s_{RC}/(s_{RC}+1)$, where $s_{RC} = j(\omega/\omega_{RC})$

SERIES R, PARALLEL C CIRCUIT

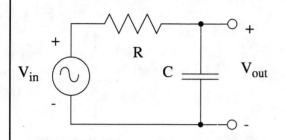

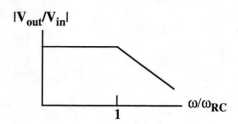

$$\left|\frac{V_{out}}{V_{in}}\right| = \frac{1}{\sqrt{(\omega RC)^2 + 1}}, \text{ or } \left|\frac{V_{out}}{V_{in}}\right| = \frac{1}{\sqrt{(\omega/\omega_{RC})^2 + 1}} \tag{9}$$

where $\omega_{RC} = 1/RC$.

In **s** notation, $V_{out}/V_{in} = 1/(s_{RC} + 1)$, where $s_{RC} = j(\omega/\omega_{RC})$.

Procedure

ADMINISTRATIVE:

1. Ask your lab instructor which machines can be used, where they are located, and when they are available.
2. Obtain an appropriate computer account if required.

RUNNING SIMULINK:

1. (See local handout on how to login.)
2. After you have logged in, type 'matlab'.
3. When the MATLAB program is running, type 'simulink'.

OPENING A NEW WINDOW:

1. In Simulink, you will need to open a workspace window in which you will draw the block diagram for your system.
   To open a new workspace window:
   a. Go to the menu 'File' with the mouse and push the left mouse button.
   b. Select 'New' with the left mouse button.
   We now have a workspace in which we can draw a new system. Note that the window is titled 'untitled'.

Part A: Gain Blocks

1. Go to the 'Sources' library icon and double click the left mouse button. This will open up the 'Sources' library.
2. Within the 'Sources' library, select the 'Sine Wave' by clicking and holding down the left mouse button. The icon should become dark if it is selected properly. Drag the icon to your blank 'untitled' workspace window and release the mouse button.
3. Close the 'Sources' library by going to the 'File' menu and selecting 'Close'.
4. Next, open the 'Linear' library.
5. Select a 'Gain' block and drag it to the 'untitled' window.
6. Double click on the 'Gain' block. This will allow you to modify the parameters of the block. Change the 'gain' parameter from the default value to 5.
7. Close the 'Linear' library.
8. Open the 'Extras' library.
9. Open the 'Display Devices' sub-library.
10. Select a 'Graph Scope' and drag it to the 'untitled' workspace window. Select another 'Graph Scope' and drag it to your workspace.
11. Close the 'Extras' library.
12. Now we are ready to connect the components of our system. To connect two blocks, simply double click and hold down the mouse button on the output port of one block. Then drag the mouse to the input port of the device you want to connect to and release the mouse button. Connect all the components as shown below:

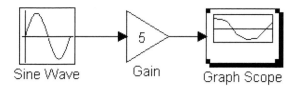

Sine Wave   Gain   Graph Scope

Figure 24.2 SIMULINK representation of gain block with sinusoidal input

Running a Simulation

1. Go to the 'Simulation' menu in the 'untitled' workspace window and click the left mouse button.
2. Select the 'Parameters' sub-menu. Change the following parameters from the default values to the values indicated:
   * Stop Time: 20
   * Minimum Step Size: 0.01
   * Maximum Step Size: 0.1
3. To start the simulation, select 'Start'
   Once the simulation has completed running, you should see a graph on the screen with two signals drawn.

   Note: You may wish to save your workspace window at this point. See section on saving workspaces.

Part B: Frequency Content of a Sine Wave

1. Create a new 'untitled' workspace window (see 'Opening a New Window').
2. Select a 'Sine Wave' (from the 'Sources' library) and drag it to the 'untitled' window.
3. Select a 'Power Spectral Density' scope (located under the 'Extras' library in the 'Display Devices' sub-library) and drag it to the 'untitled' window.
4. Double click on the 'Sine Wave' block and change the 'frequency' parameter from the default value to 5 (rad/sec).
5. Connect the blocks as shown below:

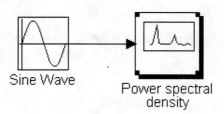

Figure 24.3 Layout of sine-wave source and spectral density indicator

6. Run a simulation (follow the steps listed in 'Running a Simulation'). After the simulation is done, the graph you will see displays the frequency content of the sine wave. You should see a spike centered at a particular frequency. This spike is the frequency domain representation of the sine-wave input.

Part C: Frequency Content of a Square Wave

1. Follow the same procedure above to construct the following system: (Note that the 'square wave' block is in the 'Sources' library).

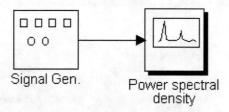

Figure 24.4 Layout of square-wave source and spectral density indicator

2. Run a simulation for the system (follow the steps in 'Running a Simulation').When your simulation is done, you should see graph showing the frequency content of the square wave.

Part D: Gain Block with Saturation

1. Set up a system consisting of a sine-wave source, a gain block, a saturation block and a spectral density indicator: Note: The saturation block limits the output at the upper and lower output limits that you choose.

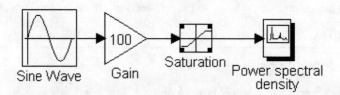

Figure 24.5 Gain block with saturation

2. Run a simulation for the system, experimenting with different gains and saturation values.
3. (Optional) Connect an audio signal source to an inexpensive (poor) amplifier and loudspeaker. Listen to the output and observe the output waveform on an oscilloscope as you increase the output signal level, causing the amplified signal to become clipped and resemble a square wave. Can you hear the change in the sound when clipping begins to occur?

Part E: Transfer Functions

1. Open a new window.
2. Create the system shown below. (Note: The block labeled 1/(s+1) is found in the 'Linear' library; set numerator to 1 and the denominator to 11. The 'Graph Scope' is found in the 'Display Devices' library.)

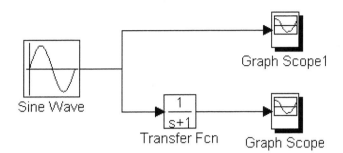

Figure 24.6 Layout for simulating a block having a given transfer function

3. Vary the frequency of the input from much less than 1 to much greater than 1; run a simulation at each frequency to see the filtering effect. (To make the two scope traces different colors, double-click on the scope and adjust its line type.) Compare the input and output signals for the various frequencies.

MISCELLANEOUS NOTES FOR SIMULINK LAB

PRINTING OUT WORKSPACE DIAGRAMS IN SIMULINK

To print out your workspace window, click and hold the mouse button over the 'File' menu in the window. Select the 'Print' option. You can send your printouts to a printer specified by your instructor.

SAVING YOUR WORKSPACE WINDOW IN SIMULINK

To save your workspace window, simply select the 'Save' option from the 'File' menu in the workspace window.

References

Simulink Manual.

Simulink

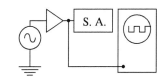

Questions for Part A

Q1. Print out your block diagram and output plots. (Ask instructor to how make printouts.)

Q2. The smaller signal on the screen is the input sine wave. The larger signal is the output of the gain block. Based on your observations, what does a gain block do?

_____

_____

_____

_____

Q3. Suppose I wanted to make the output of the gain block half as large as the input. What would I set the value of the 'Gain' block to?

_____

_____

_____

_____

Questions for Part B

Q4. Make a printout of your block diagram and output graph.

Q5. Based on your observations, what can you say about the frequency content of a sine wave? In other words, how many different frequencies does a sine wave contain?

_____

_____

_____

_____

Q6. What is the correspondence between the frequency of the sine wave we used (5 rad/sec) and the location of the spike you saw in the output graph?

_____

_____

_____

_____

Q7. If the sine wave had a frequency of 15 rad/sec instead, what would its frequency domain representation look like? (Sketch it.)

_____

_____

_____

_____

_____

Questions for Part C:

Q8. Make a printout of your block diagram and output graph.

Q9. Based on your observations, what can you say about the spectral content of a square wave?

_____

_____

_____

_____

Q10. What do you notice about the frequencies contained in the square wave?

_____

_____

_____

_____

Questions for Part D:

Q11. In what ways do the gain and saturation blocks affect the waveform and spectral content of the output?

_____

_____

Q12. (Optional) What did you observe in the amplifier/loudspeaker experiment?

_____

_____

Questions for Part E:

Q13. Based on the output graph, what kind of filter corresponds to the transfer function $1/(s+1)$? (Hint: Review your RC Filters Lab). What kind of filter would the transfer function $s/(s+1)$ correspond to? (You can change the transfer function block by double clicking on it and seting the parameter 'numerator' to [10] and the denominator to [11].

_____

_____

_____

_____

_____

Conclusions (What did you learn from this experiment?)

_____

_____

_____

Signature: _____ Date: ___/___/___ Witness: _____ Date: ___/___/___

Prelab Questions: Simulink

(Bring sheet with questions answered to your lab session)

Print your name (Last, First): _____

Q1. In simplest terms, what is a spectrum analyzer?